T0135661

CONSISTENT APPROXIMATIONS OF CONSTRAINED OPTIMAL CONTROL PROBLEMS

Vadim Azhmyakov

Fachgebiet Regelungssysteme,
Institut für Energie- und Automatisierungstechnik,
Technische Universität Berlin,
Einsteinufer 17, D-10587 Berlin, Germany.
e-mail: azhmyakov@control.tu-berlin.de

Berlin, Germany

Bibliographic information published by the Deutsche Nationalbibliothek

The Deutsche Nationalbibliothek lists this publication in the Deutsche Nationalbibliografie; detailed bibliographic data are available in the Internet at http://dnb.d-nb.de .

ISBN 978-3-8325-1584-3

Logos Verlag Berlin
Comeniushof, Gubener Str. 47,
10243 Berlin
Tel.: +49 030 42 85 10 90
Fax: +49 030 42 85 10 92
INTERNET: http://www.logos-verlag.de

*to Sabrina, Benjamin and Elisabeth
with love and thanks*

Preface

Problems of mathematical control theory are encountered in diverse areas of the exact sciences, the natural sciences and engineering. Control theory is a broad field including many topics, from theoretical studies of optimal control to very practical questions about servomechanisms and computer software. After three hundred years of evolution, optimal control theory has been formalized as a generalized extension of the calculus of variation. Developments in the vast field of optimal control are, to a great extent, motivated by real-world applications. Optimal control theory has established itself as an indispensable tool in such diverse disciplines as robotics, aeronautics, process control and economic management. In studying optimal control, we have the opportunity to invent and play variations of themes composed by such masters as Euler, Bernoulli, Lagrange, Jacobi, Newton, Hamilton, Weierstrass, Bolza, Caratheodory, Tonelli, Pontryagin, Boltyanskii, Bellman and Clarke. With the advancement of modern computer sciences, optimal control theory has become increasingly important. However, most real-world problems are much too complex to be solved exactly. Thus, computational algorithms are unavoidable in solving these applied problems. The area of numerical algorithms for optimal control has attracted the interest of many researchers from fields ranging from mathematics and engineering to biomedical and management sciences. As a result, a large body of algorithms for optimal control problems are available today in the literature. Most of these algorithms are derived for optimal control problems with constraints and involve discrete approximations. On the other hand, the main classes of extremal problems include ill-posed problems. Therefore, the use of standard optimization and discretization methods often proves to be unsuccessful for solving the ill-posed optimal control problems. The first special concept for ill-posed problems was suggested by Tikhonov in 1965. He pointed out that many optimal control problems governed by ordinary differential equations are ill-posed with respect to the convergence of the minimizing sequence. The alternative regularization method, namely, the proximal point method was introduced by Martinet in 1970. This method uses the stabilizing properties of the proximal mapping. In parallel with Tikhonov's method the proximal point algorithm is the main method for treating ill-posed problems of mathematical programming.

The aim of this book is to apply the proximal-based regularization approach to constrained optimal control problems involving ordinary differential equations. This way we incorporate in our numerical framework stable approximation procedures for a class of problems. The task of our research was to look closely at the possible computational schemes based on proximal point algorithms and to study the corresponding convergence properties of the obtained algorithms. Moreover, our book also deals with some finite-dimensional approximations of relaxed optimal control problems and with regularity conditions for a class of abstract optimal control problems. We have divided the book into seven chapters.

Chapter 1 has an introductory character. We discuss the problems associated with approximations of optimal control problems in the framework of the numerical analysis and in connection with the well- and ill-posedness concepts. We also provide an overview of the basic computational methods for optimal control problems with constraints. Chapter 2 introduces the reader to some essential aspects of proximal point method with regard to the strong convergence of the minimizing sequence. We describe a new strong convergent proximal-based algorithm proposed by the author and present some numerical schemes based on the classic and modified proximal point approaches to the convex optimization problems in Hilbert spaces. Moreover, in this chapter we examine the method of convex approximations in the abstract form. In Chapter 3 we study the optimal control problems and consider two popular approximation procedures in optimal control, namely, linearization and discretization procedures. Using the proposed techniques of the proximal point method and convex approximations, we establish a numerically stable algorithm for optimal control problems with constraints. This algorithm is based on the reduced gradients. In Chapter 4 we focus our attention on the new approximation technique for relaxed optimal control problems. We consider as before the control processes governed by ordinary differential equations in the presence of state, target and integral constraints. Chapter 6 is a short introduction to gradient-based techniques for a class of hybrid optimal control problems. Note that the optimal control of hybrid systems has become a new focus of the modern control theory. Moreover, we also briefly discuss a computational approach to optimization of controlled mechanical systems. Chapter 7 is devoted to a special class of optimal control problems in Banach spaces involving equality constraints with stable operators. We give a short introduction to the theory of stable operators, present the new analytical properties of stable and expanding operators obtained by the author and propose the new constraint qualifications for the above abstract problems. The books ends with an Appendix containing a short collection of the necessary mathematical facts from functional analysis and optimization theory.

The material presented in this monograph is the result of the author's research at the University of Greifswald and at the Technical University of Berlin. In the manuscript, standard notation is used throughout. We consider our work as a theoretical study on the numerical analysis with application to optimal control. Therefore, all computational examples presented in the manuscript are in essence illustrative examples. Every original theoretical result of the author presented in the manuscript is labeled as "Theorem" or "Lemma". For the known facts we use the term "Proposition".

This book is intended for students and professionals in applied mathematics and control engineering. The reader who wishes not only to gain access to the main results in this book but also to follow all the mathematical constructions will require a graduate-level knowledge of real and functional analysis and optimization. With this prerequisite, an advance course in computational optimization of dynamical systems can be based upon this monograph. The presented book can be used as an additional textbook for PhD student majoring in mathematical control theory and also serve as a reference for researchers in applied mathematics, control engineering and computational sciences.

The references have no pretension of completeness. They only include some works that deals with control problems through the classical optimal control theory and are directly related with the results in this book, in particular with ill-posedeness in optimization, proximal-based techniques, stable operators and approximation schemes in optimal control. We have also attempted to include a modicum of references to subjects not treated in details in this book. When possible, we have deferred to the extensive references in several recent books and papers in control theory.

During my scientific career, I had the good fortune to be able to attend many European and International workshops and conferences on state-of-the art control theory and optimization. The influence of the latter in the development of modern optimal control theory can hardly be overestimated. Several experts in control theory and optimization contributed to this monograph, and I am grateful to all of them for their support during my work. I would like to thank my colleague and friend, Prof. W.H.

Schmidt (University of Greifswald) with whom I have had inspiring and detailed discussions on many topics of optimal control theory and optimization. I am grateful to Prof. A. Kaplan (University of Trier) for his critical comments and helpful suggestions in connection with theoretical aspects of the proximal point method. The author is also grateful to all participants of the seminars and meetings held at the Institute of Mathematics and Computer Sciences (University of Greifswald) and at the Department of Control Systems (Technical University of Berlin). My warm thanks are due to Prof. L. Bittner, to Prof. B. Kugelmann (University of Greifswald) and to Prof. J. Raisch (Technical University of Berlin) for discussing and checking my ideas, theoretical results and computational shcemes. Finally, I wish to thank K. Neville (Princenton University) and T. Howell (University of Greifswald) for reading the manuscript and noting errors.

Contents

Chapter 1

Introduction and Motivation

This chapter has a tutorial character. The aim of our consideration here is to describe some of the difficulties associated with approximations of optimal control problems (OCPs for short) governed by ordinary differential equations in the framework of the numerical analysis and in connection with the well- and ill-posedness concepts. We introduce optimal control problems with target and state constraints and focus our attention primarily on the direct computational techniques and on the motivation of the stable numerical methods. We also take a quick look at the basic computational algorithms for constrained OCPs.

1.1 Well-Posedness in Optimal Control

Let us consider a proper extended-real valued functional

$$\Upsilon : \mathcal{W} \to (-\infty, \infty]$$

on a convergence space $\mathcal{W}$. The (global) minimization problem

$$\min_{w \in \mathcal{W}} \Upsilon(w) \qquad (1.1)$$

is called *Tikhonov well-posed* if and only if there exists exactly one global minimizer w^{opt} and every minimizing sequence for (1.1) converges to w^{opt}. Otherwise, (1.1) is said to be *Tikhonov ill-posed*. Note that uniqueness of the global minimizer to (1.1) does not imply well-posedness. The first special concept for ill-posed problems, suggested by Tikhonov, was proposed in [210]. In [209] Tikhonov pointed out that many OCPs (involving ordinary differential equations) are ill-posed with respect to the convergence of the minimizing sequence. Essential progress in the development of solution methods for ill-posed optimization problems was initiated by the work of Mosco [162]. A regularization method using the stabilizing properties of the *proximal mapping* (see [184, 92]) was introduced by Martinet in [152].

We now give some easy examples of well- and ill-posed OCPs governed by ordinary differential equations (see [87, 213, 225] and [18, 19]).

Example 1 *Consider the following ill posed optimal control problem*

$$\text{minimize } J(u(\cdot)) := \int_0^1 x^2(t)dt$$

subject to $\dot{x}(t) = u(t)$, a.e. on $[0, 1]$,

$x(0) = 0$,

$u(\cdot) \in \mathbb{L}^2([0, 1]), \ |u(t)| \leq 1$,

By $\mathbb{L}^2([0, 1])$ we denote the standard Lebesgue *space of all square-integrable functions $u : [0, 1] \rightarrow \mathbb{R}$. The unique optimal control is $u^{opt}(t) = 0$ a.e., however the following minimizing sequence*

$$u_r(t) = \sin(2\pi r t)$$

does not converge in the sense of $\| \cdot \|_{\mathbb{L}^2([0,1])}$. Evidently,

$$x_r(t) = \frac{1}{2\pi r}(1 - \cos(2\pi r t)), \ x_r(1) = 0$$

and

$$\lim_{r \to \infty} J(u_r(\cdot)) = \lim_{r \to \infty} \frac{3}{8\pi^2 r^2} = J(u^{opt}(\cdot)) = 0.$$

In [20] we extend the ill-posed OCP from Example 1 by the additional target condition $x(1) \leq 0$. The extended OCP is also ill-posed. It is evident that well- or ill-posedness properties of an optimization problem depend strongly on the norm under consideration.

Example 2 *The minimization of*

$$\int_0^1 (x^2(t) + u^2(t))dt$$

subject to

$$\dot{x}(t) = u(t),$$
$$x(0) = 0,$$
$$|u(t)| \leq 1 \text{ a.e.}$$

has the unique optimal control $u^{opt}(t) = 0$ a.e. on $[0, 1]$. This OCP is well-posed with respect to $\| \cdot \|_{\mathbb{L}^2([0,1])}$. If one strengthens the convergence in $\mathbb{L}^\infty(0, 1)$, then the problem becomes ill-posed. For example, the minimizing sequence

$$u_r(t) = \begin{cases} 0 & \text{if } t > 1/r, \\ s & 0 \le t \le 1/r, \end{cases}$$

where $s = $ const, does not converge in the sense of the following norm

$$\|u_r(\cdot) - u^{opt}\|_{\mathbb{L}^\infty([0,1])} = s \ \forall r \in \mathbb{N}.$$

One of the most prominent optimal control problems, namely the Fuller problem, is also an ill-posed OCP (see e.g., [104, 188]). The ill-posed optimal control problems governed by partial differential equations are of frequent occurrence. For ill-posed OCPs with partial differential equations see e.g., [188] and the references therein. For the further examples of well- and ill-posed OCPs see [225]. In our work we deal with OCPs in the presence of additional constraints. Generally the proof of the well-posedness for a constrained OCP is a very sophisticated theoretical problem. Therefore, an effective stable computational method for OCP with additional constraints contains, as a rule, a numerical procedure for regularizing possible ill-posed problems.

An OCP with different additional constraints does not need to have only one optimal solution. On the other hand, our prime interest is to study minimizing sequences generated by a concrete numerical method. In practice, we focus our attention on the convergence properties and on the consistence of a numerical method. A general theoretic investigation (in the sense of Tikhonov) of the convergence properties for every minimizing sequences is usually not necessary. Therefore, in our works [18, 19, 20, 22] we modify the standard Tikhonov's approach and use the *stability* concept (see [18], Definition 2.1, p.395 and Definition 1, Section 2.2). Needless to say that the concept of the numerical stability is weaker in comparison to the Tikhonov definition of well-posedness and is closely related to a chosen computational method. In [20, 22] we propose some modifications of the classic proximal point approach and construct numerically stable approximations for constrained OCPs. With the aid of the presented proximal-based theory usable numerical methods and optimization methods (see e.g., [199]) can also be applied to ill-posed OCPs.

It should be mentioned that in the literature alternative definitions of well-posedness in optimization occur that are based on other viewpoints. Note that well-posedness and ill-posedness concepts for variational and optimization problems are comprehensively discussed in the books of Dontchev and Zolezzi [87], Kaplan and Tichatschke [127] and Vasil'ev [213]. For the theory of *locally ill-posed problems* see [119] and [25]. An optimization problem of the type (1.1) is called *Hadamard well-posed* if there exists exactly one global minimizer w^{opt} and, roughly speaking, w^{opt} depends continuously upon the parameters of the given problem. The notation of Hadamard well-posedness in optimization reminds us of the analogous concept for boundary value problems in mathematical physics [114]. More important than the mere similarity, there are significant results, showing that many linear operator equations, or variational inequalities, are well-posed in the classical sense of Hadamard if and only if an associated minimization problem has a unique solution, which depends continuously on the papameters of the problem (see [144, 145]). There are many links between Tikhonov and Hadamard definitions of well-posedness in the optimization theory [87]. In the book [25] we consider the linkages between both well-posedness concepts in the framework of stable operators in Banach spaces (see [25], Chapter 2). This investigation has culminated in the proof of main theorems for linear ([25], Theorem 16, pp.33) and nonlinear ([25], Theorem 33 and Theorem 34, pp.77-79) differentiable stable operators. We will discuss some elements and possible applications of this theory to an abstract OCP in Chapter 7.

Finally, note that the notations of well-posedness in the optimization theory are significant as far as the numerical solution is involved. Ill-posed problems in the sense of Tikhonov or Hadamard should be handled with special care, since numerical methods will fail in general, and regularization techniques will be required.

1.2   Computational Methods

In this section we will describe, very briefly, some basic effective numerical methods for OCPs with additional target and state constraints. The survey presented here is written in view of the consistency of the corresponding numerical schemes and to give the reader an understanding of the main theoretic difficulties in connection with this. We deal with the following OCP

$$\text{minimize } J(x(\cdot), u(\cdot)) = \int_0^{t_f} f_0(t, x(t), u(t)) dt$$

subject to $\dot{x}(t) = f(t, x(t), u(t))$ a.e. on $[0, t_f]$,

$$x(0) = x_0,$$ (1.2)

$$u(t) \in U \text{ a.e. on } [0, t_f],$$

$$h_j(x(t_f)) \leq 0 \ \forall j \in I,$$

$$q(t, x(t)) \leq 0 \ \forall t \in [0, t_f],$$

where $f_0 : [0, t_f] \times \mathbb{R}^n \times \mathbb{R}^m \to \mathbb{R}$ is a continuously differentiable function,

$$f : [0, t_f] \times \mathbb{R}^n \times \mathbb{R}^m \to \mathbb{R}^n,$$

$$h_j : \mathbb{R}^n \to \mathbb{R} \text{ for } j \in I,$$

$$q : [0, t_f] \times \mathbb{R}^n \to \mathbb{R}$$

and $x_0 \in \mathbb{R}^n$ is a fixed initial state. By I we denote a finite set of index values. We assume that the functions $f, h_j(\cdot)$, $j \in I$ and $q(t, \cdot)$, $t \in [0, t_f]$ are continuously differentiable and the function f_0 is integrable. The control set U is a compact and convex subset of $\mathbb{R}^m$. We will restrict our consideration to the following control sets

$$U := \{u \in \mathbb{R}^m \ : \ b_-^i \leq u_i \leq b_+^i, \ i = 1, ..., m\},$$

where $b_-^i, b_+^i, i = 1, ..., m$ are constants. The admissible control functions $u : [0, t_f] \to \mathbb{R}^m$ are square integrable functions in time. Let

$$\mathcal{U} := \{v(\cdot) \in \mathbb{L}_m^2([0, t_f]) \ : \ v(t) \in U \text{ a.e. on } [0, t_f]\}$$

be the set of admissible control functions. Without loss of generality we suppose only one state condition, it is possible to extend our approach to optimal control problems with several state conditions. By $\mathbb{L}_m^2([0, t_f])$ we denote the standard Lebesgue space of all square-integrable functions $[0, t_f] \to \mathbb{R}^m$. In addition, we assume that for each $u(\cdot) \in \mathcal{U}$ the initial value problem

$$\dot{x}(t) = f(t, x(t), u(t)) \text{ a.e. on } [0, t_f], \ x(0) = x_0$$ (1.3)

has a unique absolutely continuous solution $x^u(\cdot)$. For some constructive uniqueness conditions see e.g., [44, 94, 120, 178] (see also Proposition

1, Section 3.1 and [20], p.3). Given an admissible control function $u(\cdot)$ the solution to the initial value problem (1.3) is an absolutely continuous function $x : [0, t_f] \to \mathbb{R}^r$. It is denoted by $x^u(\cdot)$. We assume that the problem (1.2) has an optimal solution. It is well known that the class of optimal control problems of the type (1.2) is broadly representative (see [44, 4, 120, 94]). An OCP governed by ordinary differential equations can also be considered as a problem with a terminal objective functional

$$\mathcal{J}(u(\cdot)) := \phi(x^u(t_f)),$$

where ϕ is a differentiable function. An OCP of this sort is labelled (1.2a). Recall that an OCP (1.2) with an integral functional J can be reformulated as an equivalent OCP (1.2a) with an associated terminal functional $\mathcal{J}$.

Computational methods based on the *Bellman Optimality Principle* were among the first proposed for optimal control problems [42, 63]. These methods are especially attractive when an optimal control problem is discretized a priori and the discrete version of the Bellman equation is used to solve it. Dynamic programming algorithms are widely used specifically in the engineering and operations research (see [46, 146]). For example, in [21] we applied the Bellman approach to a problem of optimal design of elastic beams. Different approximation schemes have been proposed to cope with the "curse of dimensionality" (Bellman's own phrase) with rather limited success. Moreover, the Bellman-type methods are inefficient for OCPs with constraints (communicated by H.J. Pesch).

The application of necessary conditions of optimal control theory, specifically of the *Pontryagin Maximum Principle*, yields a boundary value problem with ordinary differential equations. Clearly, the necessary optimality conditions and the corresponding boundary-value problems play an important role in optimal control computations. There is a body of work in Russian focusing on the numerical treatment of the boundary value problem for the Hamilton-Pontryagin system of differential equation (see [95, 156] and the references therein). Some of the first numerical methods for OCPs were based on the *shooting method* [58]. This technique, which was extended to a class of the problems with state constraints (*multiple shooting method*) (see [65, 66, 167]), usually guarantees very accurate solutions provided that good initial guesses for the adjoint variable $p(0)$ are available. In practice, one needs to integrate the Hamilton-Pontryagin system and then adjust the chosen initial value of the adjoint variable such that the computed $p(\cdot)$ satisfies all terminal conditions imposed on $p(t_f)$. In the view of convergence properties, the multiple shooting scheme has all advantages and disadvantages of the family of Newton methods. The general disadvantage of the *indirect numerical methods* based on the necessary optimality conditions for OCPs is the possible local nature of the obtained optimal solution. On the other hand there is also a body of work focusing on the so called Sakawa-type algorithms (see e.g., [192, 55]). This group of methods use, in fact, the idea of augmented Hamiltonian for regularizing the Pontryagin's maximizing problem. For the closely related family of Chernousko methods see also [77, 195].

It is common knowledge that an optimal control problem involving ordinary differential equations can be formulated in various ways as an optimization problem in a suitable function space (see e.g., [90, 120]). The original problem (1.2) can be expressed as an infinite-dimensional optimization problem

$$
\begin{aligned}
&\text{minimize } \tilde{J}(u(\cdot)) \\
&\text{subject to } u(\cdot) \in \mathcal{U}, \\
&\tilde{h}_j(u(\cdot)) \le 0 \; \forall j \in I, \\
&\tilde{q}(u(\cdot))(t) \le 0 \; \forall t \in [0, t_f],
\end{aligned}
\tag{1.4}
$$

with the aid of the functions

$$
\tilde{J} : \mathbb{L}^2_m([0, t_f]) \to \mathbb{R}, \; \tilde{h}_j : \mathbb{L}^2_m([0, t_f]) \to \mathbb{R}
$$

for $j \in I$ and $\tilde{q} : \mathbb{L}^2_m([0, t_f]) \to \mathbb{C}([0, t_f])$:

$$
\begin{aligned}
&\tilde{J}(u(\cdot)) := J(x^u(\cdot), u(\cdot)) = \int_0^{t_f} f_0(t, x^u(t), u(t)) dt, \\
&\tilde{h}_j(u(\cdot)) := h_j(x^u(t_f)) \; \forall j \in I, \\
&\tilde{q}(u(\cdot))(t) := q(t, x^u(t)) \; \forall t \in [0, t_f].
\end{aligned}
$$

The corresponding infinite-dimensional representation for an OCP (1.2a) with a terminal functional $\mathcal{J}$ given above is labeled (1.4a). Clearly, the given reformulation of the initial OCP (1.2) provided a basis for application of the optimization methods. Note that the *method of sequential linearizations* of Fedorenko [95], the *feasible direction algorithm* and the *function space algorithm* of Pytlak [178] substantially use the presented form (1.4) of (1.2). The complete convergence analysis for a class of methods for OCPs with state constraints is presented in [178]. In [20, 22] we also apply our proximal-type techniques to the OCPs written in the form (1.4). An effective computational algorithm based on (1.4), to be sure, includes an integrating procedure for initial value problem (1.3). Note that there is a variety of numerical algorithms for OCPs based on a possible reformulation of (1.2). One possibility is that (1.2) is represented as a minimization problem in view of the pair $(x(\cdot), u(\cdot))$. The variables $x(\cdot)$ and $u(\cdot)$ may be treated as independent variables. For instance, see the Balakrishnan's ϵ-method [38].

An optimal control problem with state constraints can also be solved by using some modern numerical algorithms of nonlinear programming. For example, the implementation of the *interior point* method is presented in [218]. The application of the *trust-region* method to optimal control is discussed in [103, 140]. In [43] the proximal point methods are used for solving stochastic and in [20, 22, 23, 27, 24, 26] for solving some classes of deterministic OCPs. For a general survey of applications of nonlinear programming to optimal control see [48]. Although there is extensive literature on computational optimal control relatively few results have been published on the application of an SQP (*Sequential Quadratic Programming*) type optimization algorithm to OCPs (see [148, 166, 212, 71] and references therein). Note that calculating of second order derivatives of the objective functional and of constraints in (1.2) can be avoided by applying an SQP-based scheme in which these derivatives are approximated by quasi-Newton formulaes. It is evident that in this case one deals with all usual disadvantages of the Newton-based methods.

Gradient-type algorithms (see [173]) are the methods based on the evaluation of the gradient for the objective functional in an OCP. The first applications of the gradient methods in optimal control are discussed in [39]. See also the work of Bryson and Denham [64] and the book of Teo/Goh/Wong [204]. The gradient of the objective functional can be computed by solving the state equations such that the obtained trajectory is then used to integrate the adjoint equations backward in time. The state and the adjoint variables are used to calculate the gradient of the functional with respect to control variable only. This is possible due to the fact that the trajectories of a control system are uniquely determined by controls. The gradient methods are widely used and are regarded as the most reliable, if not very accurate, methods for OCPs [91, 131]. There are two major sources of the inaccuracies in the gradient-type methods: errors caused by integration procedures applied to the state equations, errors introduced by the optimization algorithm. In the works [20, 22, 27, 24] of the author we also applied the evaluation of the gradient for the objective functional and the gradient method to OCPs with ordinary differential equations. In [21] we use gradients in a solution procedure for OCPs with partial differential equations. Note that the computational gradient-type methods are also studied in connection with a possible numerical approach to *relaxed* OCPs [205]. The computational aspect of the theory of β-relaxation proposed in [19, 27] also contains the technique of reduced gradients.

When solving an optimal control problem with ordinary differential equations we deal with functions and systems which, except in very special cases, are to be replaced by numerically tractable approximations. The implementation of our numerical schemes for original and relaxed control problems is based on the discrete approximations of the control set and on the finite difference approximations [20, 22, 27, 26]. Finite difference methods turn out to be a powerful tool for theoretical and practical analysis of control problems (see e.g., [160, 85, 173, 90]). Discrete approximations can be applied directly to the problem at hand or to auxiliary problems used in the solution procedure [159, 89, 178]. An example of the direct application of discrete approximations to an OCP in mechanics is considered in the paper [21] of the author. The numerical methods for optimal control problems with constraints (with the exception of the works [95, 178]) are either the methods based on the full discretizations (parametrization of state and control variables), or they are function space algorithms. The first group of methods assumed *a priori* discretization of system equations. The second group of methods is, in fact, theoretical work on the convergence of algorithms. The major drawback of some numerical schemes from the first group is the lack of the corresponding convergence analysis. This is especially true in regard to the multiple shooting method [65, 66] and to the *collocation* method of Stryk [200]. For the alternative model of collocations see e.g., [47]. Note that collocation methods are similar to the gradient algorithms applied to a priori discretized problems with the exception that gradients are not calculated with the help of adjoint equations. Instead both variables $x(\cdot)$ and $u(\cdot)$ are regarded as optimization parameters. We refer to [88, 115] for high-order schemes applied to unconstrained problems and for error estimates for discrete approximations in optimal control. Moreover, the discrete approximations to dual OCPs are studied in [116]. There are a number of results scattered in the literature on discrete approximations that are very often closely related, although apparently independent. For a survey of some recent works on computational optimal control, including discrete approximations, see [89] and [27, 24].

In our works [18, 19, 20, 22, 27, 26] we also deal with a priori discrete approximations of OCPs governed by ordinary differential equations. In [20, 22] the problem is *convex-linear*, [23, 26] introduce a new class of control systems, namely, the *convex control systems*. We attach much importance to the convergence properties of the discretizations of the initial OCP (1.2). In this case we examine not only the usual regularization technique for the above-mentioned ill-posed OCPs, but also use the proximal point approach for "regularizing" the sequence of discrete *convex approximations* to (1.2). Note that the convergence analysis of discrete approximations is of primary importance in computational optimal control (see e.g., [91, 131, 154]). In the above works we also consider the "consistent" interplay between approximations such as discretizations and linearizations of an OCP. Finally note that the easy structure of the introduced convex control systems make it possible to apply the standard gradient methods for concrete numerical computations. For this family of OCP's it is not necessary to consider the sophisticated numerical procedures like multiple shooting or collocation method of Stryk.

Chapter 2

Proximal-Type Algorithms

This chapter is devoted to the numerical schemes based on the classical and modified proximal point algorithms in Hilbert spaces. We present the results of the author in connection with the convergence analysis of a proximal-like method and establish the method of convex approximations. Moreover, we discuss some relations to the Tikhonov regularization technique.

2.1 The Basic Algorithm in Hilbert Spaces

The proximal point method, suggested by Martinet in [152] and developed by Rockafellar [184], is one of the most popular stable methods for solving nonlinear equations, convex ([13, 14]) and nonconvex ([102, 130]) optimization problems, and variational inequalities. In parallel with Tikhonov's regularization [211] the proximal point algorithm is the main method for treating ill-posed problems of mathematical programming (see, e.g., [152, 127, 129, 172]). The proximal point method is rich in applications. The first application of this method for solving the problems of determining an element $z \in X$ such that

$$0 \in T(z),$$

where X is a real Hilbert space and a multifunction $T : X \to X$ is a maximal monotone operator, was suggested by Rockafellar [184]. The proximal point method can be used for example for solving the variational inequalities [128] and for studying the asymptotic behavior at infinity solutions of evolution equations [6]. The idea of approximation by application of the proximal-like methods was extended by Benker, Hamel and Tammer [43] to optimal control problems. A great amount of works is devoted to the classical variant of the proximal point method and its various modifications (see, e.g., [185, 14, 206, 112, 113]). One can find a fairly complete review of the main results in [130, 206] and in [18, 22].

Let Z be a real Hilbert space. We examine the problem of convex minimization

$$\text{minimize } \psi(z)$$
$$\text{subject to } z \in Q, \tag{2.1}$$

where $\psi : Z \to \bar{\mathbb{R}}$ is a proper convex lower semicontinuous functional and Q is a bounded, convex, closed subset of Z. By $\bar{\mathbb{R}}$ we denote here the extended real axis, i.e. $\bar{\mathbb{R}} := \mathbb{R} \bigcup \{\infty\}$. It is evident that the above-mentioned problem of finding a zero of a maximal monotone (multivalued) operator $T \equiv \partial\psi$ is equivalent to the following problem $\psi(z) \to \min$, $z \in Z$. Here, $\partial\psi$ is the subdifferential of ψ. Let $F \subseteq Q$ be the set of optimal solutions of problem (2.1). Note that F is a convex and closed set [92]. We now introduce the *proximal mapping* [184, 130]

$$\mathcal{P}_{\psi,Q\chi} : \alpha \to \text{argmin}_{z \in Q}[\psi(z) + \frac{\chi}{2}\|z - \alpha\|^2],$$
$$\chi > 0, \ \alpha \in Z. \tag{2.2}$$

The proximal mapping possesses the following properties:

- for all $z_1, z_2 \in Z$

$$\|\mathcal{P}_{\psi,Q\chi}(z_1) - \mathcal{P}_{\psi,Q\chi}(z_2)\|^2 \leq \|z_1 - z_2\|^2 -$$
$$- \|z_1 - z_2 + \mathcal{P}_{\psi,Q\chi}(z_2) - \mathcal{P}_{\psi,Q\chi}(z_1)\|^2$$

- $\mathcal{P}_{\psi,Q\chi}(z) = z$ if and only if $z \in F$;

- the functional $\eta(\alpha) := \min_{z \in Z}[\psi(z) + \frac{\chi}{2}\|z - \alpha\|^2]$ is convex and continuously (Fréchet) differentiable on Z and

$$\nabla\eta(\alpha) = \chi(\alpha - \mathcal{P}_{\psi,Q\chi}(\alpha))$$

(differentiability of ψ is not supposed).

Note that we use all these properties consistently in proofs of our main results [18, 20, 22]. Following Rockafellar [184], Kaplan and Tichatschke [130, 129], we define the iterations of the classical proximal point method

$$z_{cl}^0 \in Q,$$
$$z_{cl}^{i+1} \approx \mathcal{P}_{\psi,Q\chi_i}(z_{cl}^i), \ i = 0, 1, \ldots, \tag{2.3}$$

where $\{\chi_i\}$ is a given sequence with $0 < \chi_i \le C < \infty$ and

$$\|z_{cl}^{i+1} - \mathcal{P}_{\psi, Q, \chi_i}(z_{cl}^i)\| \le \epsilon^i, \ i = 0, 1, \dots, \ \sum_{i=0}^{\infty} \frac{\epsilon^i}{\chi_i} < \infty. \tag{2.4}$$

For the special case $\epsilon^i = 0$, $i = 0, 1, \dots$, the method (2.3) reduces to the exact scheme

$$\bar{z}_{cl}^0 \in Q, \ \bar{z}_{cl}^{i+1} = \mathcal{P}_{\psi, Q, \chi_i}(\bar{z}_{cl}^i), \ i = 0, 1, \dots.$$

Under the condition (2.4), the sequence $\{z_{cl}^i\}$ converges in the weak topology to some element $z_{cl}^{opt} \in F$. Some alternative convergence conditions are discussed in [184, 113]. Besides, (2.4) implies the convergence of the objective values $\psi(z_{cl}^i)$ to $\psi(z_{cl}^{opt})$. In other words $\{z_{cl}^i\}$, $i = 0, 1, \dots$ is a minimizing sequence. The weak convergence of $\{z_{cl}^i\}$ may fail if instead of

$$\sum_{i=0}^{\infty} \frac{\epsilon_i}{\chi_i} < \infty \ (\text{or} \ \sum_{i=0}^{\infty} \epsilon_i < \infty)$$

in (2.4) one has only $\epsilon_i \to 0$ (see [184]). Moreover, in the specific cases the corresponding sequence of classic proximal point method converges weakly, but not strongly to a minimizing point of ψ [113]. Under some restrictive additional assumptions, Rockafellar proofs the strong convergence of $\{z_{cl}^i\}$ to a unique solution of (2.1) (see [184], Theorem 2). Note that the condition $\sigma_i \to \infty$, where

$$\sigma_i := \sum_{p=0}^{i} \frac{1}{\chi_p},$$

is the weakest condition in order to ensure that

$$\psi(z_{cl}^i) \downarrow \inf_{z \in Q} \psi(z).$$

If $\sigma_i \to \sigma < \infty$, then $\{z_{cl}^i\}$ always converges strongly:

$$\|z_{cl}^{i+r} - z_{cl}^i\| \le \sum_{p=i+1}^{i+r} \|z_{cl}^{p-1} - z_{cl}^p\| = \sum_{p=i+1}^{i+r} \frac{1}{\chi_p} \|y_p\| \le$$

$$\le \left(\sum_{p=i+1}^{i+r} \frac{1}{\chi_p} \right) \|y_{i+1}\|,$$

where

$$y_p := \chi_i(z_{cl}^{p-1} - z_{cl}^p).$$

Since $\sigma_i \to \sigma$, we see that $\{z_{cl}^i\}$ is a Cauchy sequence, and therefore converges strongly to some point z^∞, even if ψ does not have a minimizer (see [113])! It $F \neq \emptyset$, we have

$$\|z - z^\infty\| \le \sum_{p=1}^{\infty} \|z_{cl}^{p-1} - z_{cl}^p\| = \sum_{p=1}^{\infty} \frac{1}{\chi_p} \|y_p\| \le \sigma \|y_1\|,$$

and

$$\mathrm{dist}(z^\infty, F) \ge \mathrm{dist}(z, F) - \|z - z^\infty\| \ge \mathrm{dist}(z, F) - \sigma \|y_1\|,$$

where $\mathrm{dist}(\cdot, \cdot)$ is a distance function in Z. If σ is small, then

$$\mathrm{dist}(z^\infty, F) > 0,$$

and $z^\infty \notin F$. Finally, note that in [113] Güler introduced a variant of the proximal point method for (2.1). This method converges under the condition

$$\sum_{i=0}^{\infty} \frac{1}{\sqrt{\chi_i}} < \infty.$$

In [22] we apply a modification of this method to OCPs with constraints.

2.2 Convergence Properties of Algorithms

From the view-point of the numerical analysis and constructive computational methods for optimization problems (for example, for optimal control problems) the strong convergence of a minimizing sequence is of practical significance. Numerical methods for optimal control problems based on the minimizing sequence of controls [95, 178] are especially attractive when we have the strongly convergent minimizing sequence of control functions. We understand the "consistence" and "numerical stability" of an approximating technique for (2.1) in the sense of the following definition (see [18], Definition 2.1, p.395)

Definition 1 *A method for solving the problem* (2.1) *is called stable if the associated minimizing sequence* z^k, $k = 0, 1, ...$ *converges strongly to some element* $z^{opt} \in F$, *i.e. if from*

$$z^k \in Q, \ \psi(z^k) \to \psi(z^{opt}) \equiv \min_{z \in Q} \psi(z),$$

follows $z^k \to z_{opt}$ *strongly for all* $z^0 \in Q$.

Some stability criteria in the sense of the Definition 1 are also considered by Polyak [172]. Note that the approach of Polak is based on the technics of the Lyapunov functions. Now we suggest the normality definition in the sense of Vasil'ev [213] (see also [18], Definition 2.2, p.395)

Definition 2 *Let* $Q_\Omega \subseteq Q$ *and* $F_\Omega := Q_\Omega \cap F \neq \emptyset$. *An optimal solution* $z^{opt} \in F_\Omega$ *of* (2.1) *is called normal with respect to some function* Ω : $Q_\Omega \to \bar{\mathbb{R}}$ *if*

$$\min_{F_\Omega} \Omega(z) = \Omega(z^{opt}).$$

The classical proximal mapping can be used for creating a strong convergent minimizing sequence for problem (2.1). The questions of the *strong convergence* of proximal-like methods were considered for example by Bakushinskii [37], by Lemaire [141, 142], Moudafi [163] and by Azhmyakov and Schmidt [18]. Related results are proved by Solodov and Svaiter in [198]. The strong convergence of an approximation method which combines the method of Tikhonov and proximal point algorithm is examined by Lehdili and Moudafi [139]. In our paper [18] we examine the strong convergence of a new proximal-based method for the problem (2.1). Using the classical proximal mapping $\mathcal{P}_{\psi,Q,\chi}$, we can construct the iterative procedure ([18], p.397)

$$\begin{aligned} z^i &= a^i z^0 + (1 - a^i)\mathcal{P}_{\psi,Q,\chi_i}(z^i), \\ i &= 0, 1, ... , \end{aligned} \tag{2.5}$$

where $\{\chi_i\}$ is a given sequence with $0 < \chi_i \le C < \infty$ and

$$a^0 = 1, \ 0 < a^i < 1, \ \forall i = 1, 2, ... , \ a^i \downarrow 0.$$

The sophisticated fixed point problem (2.5) can be approximated by the following iterative scheme ([18], pp.399-400)

$$z^i_{r+1} = a^i z^0 + (1 - a^i) \mathcal{P}_{\psi, Q, \chi_i}(z^i_r),$$
$$z^i_0 = z^{i-1}_{N(i-1)}, \ z^0_0 = z^0 \in Q, \tag{2.6}$$
$$i = 0, 1, \dots, \ r = 0, 1, \dots N(i) - 1,$$

where $N(i) \in \mathbb{N}$, $N(0) = 1$ and $\{a^i\}$ is the sequence given above. Let us now formulate the main results of [18] (see [18], Theorem 3.1, p.398; Theorem 4.1, p.401; Theorem 4.3, p.404).

Theorem 1 *The sequence $\{z^i\}$ of the solutions of the equations (2.5) converges strongly to some solution $z^{opt} \in F$ of problem (2.1). Let*

$$\Omega(z) := \|z - z^0\|, \ z \in F, \ z^0 \in Q.$$

Then the element $z^{opt} \in F$ is a unique normal solution of problem (2.1) with respect to the function $\Omega(\cdot)$.

Theorem 2 *Suppose the sequences $\{N(i)\}$, $\{a^i\}$, $\{\chi_i\}$ in (2.6) satisfy the following conditions*

$$\sum_{i=0}^{\infty} \frac{a^i N(i)}{\chi_i} < \infty, \ 0 < \chi_i \leq C < \infty.$$

Then the sequence

$$\{z^i_{r+1}\}, \ i = 0, 1, \dots, \ r = 0, \dots, N(i) - 1,$$

generated by (2.6) is a minimizing sequence. This sequence converges weakly to an element of the set F.

Theorem 3 *Suppose the sequence $\{N(i)\}$ satisfies the condition*

$$\log_{b_i}\left(\frac{a^i}{qa^{i+1}}\right) + N(i) \log_{b_i}\left(\frac{1}{1 - a^i}\right) \leq N(i + 1),$$
$$i = 1, 2, \dots,$$
$$0 < q < 1, \ N(1) \in \mathbb{N},$$

where

$$b_i := \frac{1}{1 - a^{i+1}}, \ i = 1, 2, \dots .$$

Assume that $\{z^i\}$ generated by the equation (2.5) converges strongly to some element $z^{opt} \in F$. Then the minimizing sequence $\{z^i_{N(i)}\}$ generated by the method (2.6) converges strongly to the same element $z^{opt} \in F$.

The presented theorems provide a possible answer to Rockafellar's question: "the question of whether the weak convergence established by Martinet can be improved to strong convergence thus remains open" (see [184], p.879). Note that the answer is known to be affirmative in the case of a quadratic functional ψ. This follows from the Krasnoselskii Theorem [133], as has been noted by Kryanev [135]. The proofs of the formulated theorems are based on the Browder Theorem [62, 197] and on the analytic facts for *sunny retracts* developed by Shioji and Takahashi [197] (see also [137]). Moreover, we use some standard techniques of nonlinear analysis. It is well known that a minimizing sequence generated by Tikhonov method converges to the set of normal solutions of problem (2.1) (in the sense of Definition 2) [211, 213]. In the general case the sequence $\{z^i_{cl}\}$ does not possess this property. Note that the concept of normal solutions plays an important role in operations research [213].

In our case we deal with a proximal-like method (2.5)-(2.6) that converges to a normal solution $z^{opt} \in F$ of problem (2.1). Moreover, the proposed iterative procedures are "dissipative". Let

$$z^0, \mathcal{P}_{\psi, Q, \chi_i}(z^i) \in \mathcal{W}^i_F,$$

where $\mathcal{W}^i_F$ is a convex neighborhood of the bounded, convex, closed set F. One can prove that $z^i \in \mathcal{W}^i_F$. Therefore, we call the method (2.5) "dissipative". On the other hand, the iterative schemes described in [37] and in [139] do not possess (in general) these properties. In [18] (Theorem 3.2, pp.398-399) we also extend the strongly convergent proximal point approach to the convex optimization problems in uniformly convex and uniformly smooth real Banach spaces (see e.g., [132]).

Theorem 4 *Let $\{y^i\}$ be the sequence of solutions of the equations*

$$y^i = \tilde{a}^i y^0 + (1 - \tilde{a}^i) \frac{1}{i+1} \sum_{j=0}^{i} \mathcal{P}^j_{\tilde{\psi}, K, \chi_i}(y^i), \ y^0 \in \tilde{Q},$$

where $i = 0, 1, \ldots \tilde{Q}$ is a bounded, convex, closed subset of a uniformly convex and uniformly smooth real Banach space $\tilde{Z}$,

$$\tilde{\psi} : \tilde{Z} \to \bar{\mathbb{R}}$$

is a proper convex lower semicontinuous functional and $\{\tilde{a}^i\}$ is a real sequence such that $\tilde{a}^0 = 1$, $0 < \tilde{a}^i \leq 1$ and $\tilde{a}^i \to 0$ and

$$\mathcal{P}_{\tilde{\psi}, K, \chi_i}(y^i) := \operatorname{argmin}_{y \in \tilde{Q}} [\tilde{\psi}(y) + \frac{\chi_i}{2} \|y - y^i\|^2].$$

Then $\{y^i\}$ converges strongly to some solution of the following problem

$$\text{minimize } \tilde{\psi}(y)$$

$$\text{subject to } y \in \tilde{Q}.$$

The classical proximal point method (2.3)-(2.4) is a variant of the general proximal-like methods using *Bregman functions* and *Bregman distances* in a real Hilbert space ([67]) or in a reflexive real Banach space ([70, 121, 68, 69]). The proximal-like method in Banach spaces was analyzed for the case of a quadratic Bregman function [3] and for the case of a general Bregman function ([70, 40, 41]). The approach proposed in [18] can also be used for the construction of the corresponding strongly convergent variant of the general proximal-like method with Bregman distances considered by Burachik and Iusem [67]. However, even in the case of real Hilbert space, the square of the norm leads always to simpler numerical algorithms. The scheme with a nonquadratic Bregman function has been proposed mainly with penalization purposes. We refer to [75, 121] for details.

In [18] we also derive the useful estimates for the presented method (2.6) ([18], Lemma 4.1, pp.400-401)

$$\|z_0^i - \mathcal{P}_{\psi, Q, \chi_i}(z_0^i)\| \leq \frac{2}{\sqrt{3\chi_i}} \sqrt{\psi(z_0^i) - \psi(\mathcal{P}_{\psi, Q, \chi_i}(z_0^i))}. \qquad (2.7)$$

Note that the following estimate is also correct for an arbitrary point $z \in Q$

$$\sqrt{\frac{2}{\chi}(\psi(z) - \psi(\mathcal{P}_{\psi, Q, \chi}(z)))} \geq \|z - \mathcal{P}_{\psi, Q, \chi}(z)\|.$$

In the case of a Lipschitz continuous functional ψ one can use (2.7) for creating the general estimate ([18], Theorem 4.2, pp.402-403).

Theorem 5 *Let $\{z^i\}$ be the sequence generated by the equation (2.5). Assume that $\{z^i\}$ converges strongly to some element $z^{opt} \in F$ and the functional ψ is* Lipschitz *continuous with* Lipschitz *constant L. Let $\{z^i_{r+1}\}$ be the sequence generated by (2.6). Then the following estimation*

$$\|z^{opt} - z^i_{N(i)}\| \le \|z^{opt} - z^i\| + (1 - a^i)^{N(i)} \mathrm{diam}Q +$$
$$+ \frac{2(1 - a^i)^{N(i)}}{a^i \sqrt{3\chi_i}} \sqrt{|\psi(z^i_0) - \psi(z^i_1)| + a^i L \mathrm{diam}Q},$$

where

$$\|z^{opt} - z^i\| \to 0,$$

holds.

Example 3 *We now apply our method (2.6) to* OCP *from Example 1 and compute the approximate optimal control $\tilde{u}(\cdot)$. Clearly, we have*

$$Z = \mathbb{L}^2(0, 1)$$

and

$$J(u(\cdot)) = \int_0^1 \Big(\int_0^t u(\tau)d\tau \Big)^2 dt.$$

It is easy to see that the functional J here is lower semicontinuous and proper convex. The set of all admissible control functions is bounded, convex and closed subset of the Hilbert *space $\mathbb{L}^2(0, 1)$. The computed optimal control $\tilde{u}(\cdot)$ satisfies the inequality*

$$\|u^* - \tilde{u}\|_{\mathbb{L}^2} \le \delta = 10^{-5}.$$

We have considered seven i-steps. Therefore, we have

$$\sum_{i=0}^6 N(i) = 60$$

iterations of the algorithm. The approximation $\tilde{u}(\cdot)$ of the optimal control has the following bang-bang *type*

$$\tilde{u}(t) = \begin{cases} h & \textit{if } t \in [t_{2j-1}, t_{2j}), \\ -h & \textit{otherwise}, \end{cases}$$

where h is a positive constant h < δ and

$$[t_{2j-1}, t_{2j}) \subset [0, 1], \ j = 1, ..., M \ (M \in \mathbb{N})$$

are some equidistant half-open intervals. The OCP from Example 1 has been solved (with the preassigned exactness δ = 10^{-5}) by application of the Bakushinskii method [37]. In this case one has used more iterations.

The implementation of the algorithm (2.6) was carried out, using the "Numerical Recipes in C" package [177] and the author's program written in C.

2.3   Convex Approximations

In the papers [20] and [22] the author proposes a new stable method of sequential discrete approximations for OCPs with constraints. For this purpose, we first analyze the related abstract discretizations of convex minimization problems (2.1) in real Hilbert spaces. Let $\{Z_N\}$, $N \in \mathbb{N}$ be a sequence of subspaces of Z such that

$$Z \supseteq ... \supseteq Z_{N+1} \supseteq Z_N, ..., \supseteq Z_1.$$

The norm $\| \cdot \|_{Z_N}$ of a subspace Z_N is induced by the norm $\| \cdot \|_Z$ of the Hilbert space Z. In parallel with (2.1) one can consider the sequence of the minimization problems

$$\begin{aligned} &\text{minimize } \psi(z)\\ &\text{subject to } z \in Q_N, \end{aligned} \qquad (2.8)$$

where

$$\{Q_N\}, \ Q_N \subset Z_N$$

is a sequence of bounded, convex, closed subsets of Z_N

$$Q_N = Z_N \bigcap Q, \ N \in \mathbb{N}.$$

In fact we deal with some restrictions of the function $\psi(\cdot)$ on Z_N, however, we use the same notation $\psi(\cdot)$. Denote

$$\psi^{opt} := \inf_{z \in Q} \psi(z),$$

$$\psi^{opt,N} := \inf_{z \in Q_N} \psi(z).$$

We introduce the following concept ([20], Definition 1, p.7)

Definition 3 *The sequence of problems (2.8) is called an approximating sequence for (2.1) if*

$$\lim_{N \to \infty} \psi^{N,opt} = \psi^{opt}.$$

Evidently, an approximating sequence corresponds a sequence of possible consistent discretizations of (2.1). These discretizations are usually finite-dimensional. Let $z_N^{opt} \in Q_N$ be an optimal solution of (2.8). At first, we extended a result of Vasil'ev [213] (see [20], Theorem 1, pp.7-8).

Theorem 6 *Assume that for all $z^{opt} \in F$ there exists a mapping*

$$P_N : Z \to Z_N, \ N \in \mathbb{N}$$

such that

$$\psi(P_N(z^{opt})) - \psi(z^{opt}) \le \gamma_N,$$

where $\lim_{N \to \infty} \gamma_N = 0$. Then (2.8) is an approximating sequence for problem (2.1).

We call the mapping from Theorem 6 a *Vasil'ev mapping*. Let

$$\mathcal{P}_{\psi,Q_N,K_N}^s(z_N) := \underbrace{(\mathcal{P}_{\psi,Q_N,K_N} \circ \dots \circ \mathcal{P}_{\psi,Q_N,K_N})}_{s}(z_N), \ s \in \mathbb{N}.$$

We now apply the classic proximal mapping and define the sequence of approximations ([20], p.8)

$$
\begin{aligned}
&z_1 \in Q_1, \\
&z_{N+1} = \mathcal{P}_{\psi,Q_{N+1},K_{N+1}}(\mathcal{P}_{\psi,Q_N,K_N}^s(z_N)), \\
&\infty > \tilde{K} > K_N \downarrow K > 0, \\
&s \in \mathbb{N}, \ N = 1, 2, \dots,
\end{aligned}
\tag{2.9}
$$

where $\tilde{K}, K$ are constants and

$$\mathcal{P}_{\psi,Q_{N+1},K_{N+1}}(\mathcal{P}_{\psi,Q_N,K_N}^s(z_N)) := \text{Argmin}_{z \in Q_{N+1}}[\psi(z) +$$
$$+ \frac{K_{N+1}}{2}\|z - \mathcal{P}_{\psi,Q_N,K_N}^s(z_N)\|_{Z_{N+1}}^2], \ z_N \in Q_N.$$

We call the proximal-based method (2.9) "Sequential Proximal-Based Method". Using some estimates for $\mathcal{P}^s_{\psi,Q_N,K_N}(\cdot)$ ([20], Lemma 2, pp.8-9), we proved the following results ([20], Theorem 2, pp.9-10)

Theorem 7 *Let $\{z_N\}$ be the sequence generated by the Sequential Proximal-Based Method (2.9). Assume that for all $z^{opt} \in F$ there exists a (Vasil'ev) mapping $P_N : Z \to Z_N$, $N \in \mathbb{N}$ such that*

$$\psi(P_N(z^{opt})) - \psi(z^{opt}) \leq \gamma_N,$$
$$\lim_{N \to \infty} \gamma_N = 0.$$

Then

$$\lim_{N \to \infty} \lim_{s \to \infty} \psi(z_N) = \psi^{opt}$$

and

$$\lim_{N \to \infty} \lim_{s \to \infty} \|z_N - \mathcal{P}^s_{\psi,Q_N,K_N}(z_N)\|_Z = 0.$$

Our next result is a consequence of Theorem 7 ([20], Corollary 1, pp.10-11).

Theorem 8 *Let $\{z_N\}$ be the sequence generated by the Sequential Proximal-Based Method (2.9). Assume that for all $z^{opt} \in F$ there exists a (Vasil'ev) mapping $P_N : Z \to Z_N$, $N \in \mathbb{N}$ such that*

$$\psi(P_N(z^{opt})) - \psi(z^{opt}) \leq \gamma_N,$$
$$\lim_{N \to \infty} \gamma_N = 0.$$

Then there exists a sequence of numbers $\{s_i\}$, $s_i \in \mathbb{N}$ such that

$$\lim_{N \to \infty} \lim_{s_i \to \infty} \|z_{N+1} - z_N^{opt}\|_Z = 0.$$

We now discuss an alternative approximation scheme, namely, so called "Proximal-Penalty Method" proposed in [22]. Using the convex approximations given above and a sequence of the auxiliary problems (2.8), we define the following sequence of approximations ([22], p.19)

$$\zeta_1 \in Q_1, \quad \zeta_{N+1} = \mathcal{P}_{\psi,Q_{N+1},K_N}(\zeta_N), \quad (2.10)$$
$$N \in \mathbb{N},$$

where

$$\mathcal{P}_{\psi,Q_{N+1},K_N}(\zeta_N) := \text{Argmin}_{\zeta \in Q_{N+1}} [\psi(\zeta) + \frac{K_N}{2} \|\zeta - \zeta_N\|_{Z_{N+1}}^2],$$

$$\zeta_N \in Q_N, \ K_N > 0,$$

and

$$\sum_{N=1}^{\infty} \frac{1}{\sqrt{K_N}} < \infty.$$

Note that the Proximal-Penalty Method (2.10) is free of the *s*-steps iterations $\mathcal{P}_{\psi,Q_N,K_N}^s$ of the classical proximal mapping used in (2.9). In contrast to the Sequential Proximal-Based Method (2.9), the proposed Proximal-Penalty Method (2.10) generates at once the $(N + 1)$-iteration over the set Q_{N+1}. The structure of the algorithm (2.10) make it possible to obtain the strengthened estimate in the case of a Lipschitz continuous functional ψ (see [22], Lemma 3.1 - Lemma 3.3, pp.20-21). Moreover, the sequence generated by (2.10) is a strongly convergent minimizing sequence for the initial problem (2.1) ([22], Theorem 3.2)

Theorem 9 *Let* $\{\zeta_N\}$ *be the sequence generated by the Proximal-Penalty Method* (2.10). *Assume that for all* $z^{opt} \in F$ *there exists a* (Vasil'ev) *mapping* $P_N : Z \rightarrow Z_N, \ N \in \mathbb{N}$ *such that*

$$\psi(P_N(\zeta^{opt})) - \psi(z^{opt}) \leq \gamma_N,$$

$$\lim_{N \rightarrow \infty} \gamma_N = 0.$$

Then $\{\zeta_N\}$ *is a minimizing sequence for the initial problem* (2.1). *This sequence converges strongly to an optimal solution* $z^{opt} \in F$ *of problem* (2.1)

$$\lim_{N \rightarrow \infty} \|\zeta_N - z^{opt}\|_Z = 0.$$

The result of Theorem 9 can be interpreted in terms of Definition 3. The sequence of the auxiliary problems

$$\text{minimize } \psi(z) + \frac{K_N}{2} \|z - z_N\|_{Z_{N+1}}^2 \tag{2.11}$$

$$\text{subject to } z \in Q_{N+1},$$

where

$$K_N > 0, \ \sum_{N=1}^{\infty} \frac{1}{\sqrt{K_N}} < \infty,$$

is an approximating sequence for the initial problem (2.1). The corresponding sequence of solutions to (2.11) is a strongly convergent minimizing sequence for the initial problem (2.1).

Chapter 3

Linear and Discrete Approximations

In the preceding chapter we have developed a new computational approach to convex optimization problems in a real Hilbert space based on the convex approximations and on a proximal-type method. To make a step forward in the study of OCPs we will discuss some numerical aspects of the popular approximations procedures in optimal control, namely, of the linearization and discretization procedures. Linearization and discrete approximation techniques have been long time recognized as a powerful tool for solving OCPs governed by ordinary differential equations. In this chapter we apply both of these approximation techniques and present a numerically stable algorithm for OCPs of the type (1.2).

3.1   Stable Solutions to Constrained OCPs

The discretization of optimization problems is an approximation procedure whose accuracy, as it is typical in numerical analysis, depends on the regularity properties of the solutions. Discrete approximations can be applied directly to the problem at hand or to auxiliary problems used in the solution procedure. In [20, 22] we mainly focused our attention on the application of a regularization method, namely, on the application of a proximal-based method to the discrete approximations of OCPs (1.2). We are particulary interested in studying the convergence of these discrete approximations. Note that one of the first application of the proximal-point method to optimal control problems is presented in [43]. We consider the first order Riemann-Euler approximations [85, 89, 90]. Some alternative discretization procedures are described in [214]. The application of the proximal-based regularization methods makes it possible to obtain the convergence results also in the case of relatively easy discretization schemes for differential equations, namely, for Riemann-Euler approximations. Clearly, the same method can be combined with other integrating procedures.

In [20] and [22] we deal with the OCPs (1.2) and (1.2a) on the time-interval $[0, 1]$. Let $x_0 = 0$. The ensuing analysis is restricted to proper convex on $\mathbb{R}^n \times \mathbb{R}^m$ function $f_0(t, \cdot, \cdot)$, $t \in [0, 1]$. In parallel with (1.3) we examine the corresponding linearized control system

$$\dot{y}(t) = f_x(t, x^u(t), u(t))y(t) + f_u(t, x^u(t), u(t))d(t),$$
$$y(0) = 0, \tag{3.1}$$

where $u(\cdot) \in \mathcal{U}$, $d(\cdot) \in \mathbb{L}_m^2([0, 1])$. Evidently, (3.1) is a linearization of the initial control system (1.3) around $(x^u(\cdot), u(\cdot))$. We assume that for each function $u(\cdot) \in \mathcal{U}$ and $d(\cdot) \in \mathbb{L}_m^2([0, 1])$ the linear initial value problem (3.1) has a unique solution (see e.g., [4, 178]). The solution to (3.1), which depends on controls

$$u(\cdot) \in \mathcal{U}, \ d(\cdot) \in \mathbb{L}_m^2([0, 1]),$$

is written $y^{u,d}(\cdot)$. Evidently,

$$y^{u,d}(t) = \int_0^t \Phi(t, s)f_u(s, x^u(s), u(s))d(s)ds \text{ a.e. } t \in [0, 1],$$

where $\Phi(\cdot, s)$ is the fundamental solution matrix prescribed by the initial value problem

$$\frac{\partial}{\partial t}\Phi(t, s) = f_x(t, x^u(t), u(t))\Phi(t, s) \text{ a.e. } t \in [0, 1],$$
$$\Phi(s, s) = E, \ s \in [0, 1].$$

Both of the above cited works of the author use the representation of the linearized problem (1.2) in the form of the corresponding *convex-linear* optimization problem

$$\begin{aligned}
&\text{minimize } \tilde{J}(d(\cdot)) \\
&\text{subject to } d(\cdot) \in \mathcal{U} - u(\cdot), \\
&\tilde{h}_j(u(\cdot)) + \langle \nabla \tilde{h}_j(u(\cdot)), d(\cdot) \rangle \leq 0 \ \forall j \in I, \\
&\tilde{q}(u(\cdot))(t) + \langle \nabla \tilde{q}(u(\cdot))(t), d(\cdot) \rangle \leq 0 \ \forall t \in [0, 1],
\end{aligned} \tag{3.2}$$

where $u(\cdot)$ is a fixed element of $\mathcal{U}$ and

$$\tilde{J}(d(\cdot)) := J(y^{u,d}(\cdot), u(\cdot) + d(\cdot)) = \int_0^1 f_0(t, y^{u,d}(t), u(t) + d(t))dt,$$

$$\langle \nabla \tilde{h}_j(u(\cdot)), d(\cdot) \rangle := (h_j)_x(x^u(1))y^{u,d}(1) \text{ for } j \in I,$$

$$\langle \nabla \tilde{q}(u(\cdot))(t), d(\cdot) \rangle := q_x(t, x^u(t))y^{u,d}(t) \text{ for } t \in [0, 1].$$

Clearly, $\hat{J}$ is a convex functional. The linear approximation (3.2) determines the first step of our computational algorithm. Moreover, the similar numerical approximations have found wide use in optimal control [204, 178]. Let us now present the consistency result (see [178]).

Proposition 1 *Assume that:*

(i) $f(t, \cdot, \cdot)$ is differentiable,

(ii) f, f_x, f_u are continuous and there exists a constant $S < \infty$ such that
$$\|f_x(t, x, u)\| \leq S \quad \forall (t, x, u) \in [0, 1] \times \mathbb{R}^n \times U.$$

Then under the conditions given in Section 1.2 *there exists a function*
$$o_1 : (0, \infty) \to (0, \infty)$$
such that $s^{-1}o_1(s) \to 0$, as $s \downarrow 0$ and
$$\|x^{u+d}(\cdot) - (x^u(\cdot) + y^{u,d}(\cdot))\|_{\mathbb{L}_m^\infty([0,1])} \leq o_1(\|d\|_{\mathbb{L}_m^2([0,1])}),$$

for all $u(\cdot) \in \mathcal{U}$ and $d(\cdot) \in \mathbb{L}_m^2([0, 1])$.

In the case of a consistent starting point $(x^u(\cdot), u(\cdot))$, the convex-linear problem (3.2) approximates the initial OCP (1.2). Following the suggested linearization step, we consider discrete schemes for (3.2) (see [20, 22]). Let N be a sufficiently large positive integer number and
$$G_N := \{t_0 = 0, t_1, ..., t_N = 1\}$$
be a (possible nonequidistant) partition of $[0, 1]$ with
$$\max_{0 \leq k \leq N-1} |t_{k+1} - t_k| \leq \xi_N.$$

We assume that $\lim_{N \to \infty} \xi_N = 0$. Define
$$\Delta t_{k+1} := t_{k+1} - t_k, \quad k = 0, ..., N - 1$$
and consider the following finite-dimensional optimization problem

$$\begin{aligned}
& \text{minimize } \tilde{J}(d_N(\cdot)) \\
& \text{subject to } d_N(\cdot) \in \mathcal{U}_N - u_N(\cdot), \\
& \tilde{h}_j(u_N(\cdot)) + \langle \nabla \tilde{h}_j(u_N(\cdot)), d_N(\cdot) \rangle \leq 0 \; \forall j \in I, \\
& \tilde{q}(u_N(\cdot))(t) + \langle \nabla \tilde{q}(u_N(\cdot))(t), d_N(\cdot) \rangle \leq 0 \; \forall t \in [0, 1],
\end{aligned} \quad (3.3)$$

where

$$\tilde{J}(d_N(\cdot)) := \sum_{k=0}^{N-1} f_0(t_k, y_N(t_k), u^k + d^k)\Delta t_{k+1},$$

$$\mathcal{U}_N := \{v_N(\cdot) \in \mathbb{L}_m^{2,N}(G_N) \ : \ v_N(t) \in U \ \forall t \in G_N\},$$

$$y_N(t_{k+1}) = y_N(t_k) + \Delta t_{k+1}(f_x(t_k, x^u(t_k), u^k)y_N(t_k) +$$

$$+ f_u(t_k, x^u(t_k), u^k)d^k), \ y_N(t_0) = 0,$$

and

$$u_N(t) := \sum_{k=0}^{N-1} \tilde{\phi}_k(t)u^k, \ u^k = u(t_k), \ d_N(t) := \sum_{k=0}^{N-1} \tilde{\phi}_k(t)d^k, \ d^k = d(t_k),$$

$$t \in [0,1], \ k = 0, 1, ...N-1 \ , \ \tilde{\phi}_k(t) := \begin{cases} 1 & \text{if } t \in [t_k, t_{k+1}[, \\ 0 & \text{otherwise.} \end{cases}$$

Clearly, $u^k \in U$ and $u^k + d^k \in U$. In effect, we deal with the finite-dimensional Hilbert space $\mathbb{L}_m^{2,N}(G_N)$ of the piecewise constant control functions $u_N(\cdot)$. The scalar product and the norm in the space $\mathbb{L}_m^{2,N}(G_N)$ are defined as follows

$$\langle u_N(\cdot), v_N(\cdot)\rangle_{\mathbb{L}_m^{2,N}(G_N)} := \sum_{k=0}^{N-1} \langle u^k, v^k\rangle_{\mathbb{R}^m},$$

$$\|u_N(\cdot)\|_{\mathbb{L}_m^{2,N}(G_N)} := (\langle u_N(\cdot), u_N(\cdot)\rangle_{\mathbb{L}_m^{2,N}(G_N)})^{1/2} = (\sum_{k=0}^{N-1} \|u^k\|_{\mathbb{R}^m}^2)^{1/2}.$$

The space $\mathbb{L}_m^{2,N}(G_N)$ is in one-to-one correspondence with the Euclidean space $\mathbb{R}^{mN}$. The Hilbert space $\mathbb{L}_m^2([0,1])$ and the set $\mathcal{U}$ are replaced by the Hilbert space $\mathbb{L}_m^{2,N}(G_N)$ and by $\mathcal{U}_N$, respectively. Evidently, we have a restriction of the function $\hat{J}(\cdot)$ on $\mathbb{L}_m^{2,N}(G_N)$.

The discrete convex-linear minimization problem (3.3) approximates the continuous problem (3.2). Note that (3.3) has an optimal solution [92]. In [20, 22] we mainly focused our attention on the application of a proximal-based method to the discrete convex-linear optimization problem (3.3) and on the convergence of the corresponding discrete approximation schemes. We apply the convex approximations given in Section 2.3 to the convex-linear problem (3.2). Our first algorithm is based on the Sequential Proximal-Based Method (2.9). We will consider the first order method for solving the auxiliary problems but similar results can also be obtained for a second order scheme. Let $d^{opt}(\cdot)$ be an optimal solution of (3.3). We introduce the sequence of spaces

$$\mathbb{L}_m^2([0, 1]) \supseteq ... \supseteq \mathbb{L}_m^{2,N+1}(G_N) \supseteq \mathbb{L}_m^{2,N}(G_N) \supseteq ... \supseteq \mathbb{L}_m^{2,1}(G_N),$$

and the sequence of sets

$$\mathcal{U} \supseteq ... \supseteq \mathcal{U}_{N+1} \supseteq \mathcal{U}_N \supseteq ... \supseteq \mathcal{U}_1, \ N \in \mathbb{N}$$

such that $\mathcal{U}_N := \mathcal{U} \cap \mathbb{L}_m^{2,N}(G_N)$. Evidently, a sequence of (3.3) is a sequence of minimization problems of the type (2.8). Let us consider the mapping

$$P_N : \mathbb{L}_m^2([0, 1]) \to \mathbb{L}_m^{2,N}(G_N), \ N \in \mathbb{N}$$

defined as follows

$$P_N(d^{opt}(\cdot)) = (d^0, ..., d^{N-1})^T, \ d^k := \frac{1}{\Delta t_{k+1}} \int_{t_k}^{t_{k+1}} d^{opt}(t)dt,$$

$$k = 0, ..., N - 1.$$

In [22] we prove that P_N is a Vasil'ev mapping for an OCP of the type (1.2) (Theorem 3, pp.13-14). We now establish two important properties of P_N (see [20], Lemma 3, Lemma 4, pp.11-12).

Theorem 10 *For every $N \in \mathbb{N}$ and for every solution $d^{opt}(\cdot)$ of problem (3.3),*

$$P_N(d^{opt}(\cdot)) \in \mathcal{U}_N - u_N(\cdot).$$

Fix $N \in \mathbb{N}$, $u_N(\cdot) \in \mathcal{U}_N$ and consider the mapping

$$P_{N+M}(d^{opt}(\cdot)), \ M = 1, 2, ...$$

for the corresponding partition $G_{N+M} := \{t_0 = 0, t_1, ..., t_{N+M} = 1\}$ of $[0, 1]$ such that $G_N \subset G_{N+M}$. Then

$$P_{N+M}(d^{opt}(\cdot)) \in \mathcal{U}_{N+M} - u_N(\cdot), \ M = 1, 2, ...$$

We now define a discrete approximation of the type (3.3) for the convex-linear problem (3.2), namely, the problem

$$\text{minimize } \tilde{J}(d_{N+M}(\cdot))$$
$$\text{subject to } d_{N+M}(\cdot) \in \mathcal{U}_{N+M} - u_N(\cdot),$$
$$\tilde{h}_j(u_N(\cdot)) + \langle \nabla \tilde{h}_j(u_N(\cdot)), d_{N+M}(\cdot) \rangle \leq 0 \; \forall j \in I, \quad (3.4)$$
$$\tilde{q}(u_N(\cdot))(t) + \langle \nabla \tilde{q}(u_N(\cdot))(t), d_{N+M}(\cdot) \rangle \leq 0 \; \forall t \in [0,1],$$
$$M = 1, 2, \dots,$$

where

$$\tilde{J}(d_{N+M}(\cdot)) = \sum_{k=0}^{N+M-1} f_0(t_k, y_{N+M}(t_k), u_N(t_k) + d_{N+M}(t_k))\Delta t_{k+1},$$

$$y_{N+M}(t_{k+1}) = y_{N+M}(t_k) + \Delta t_{k+1}(f_x(t_k, x^{u_N}(t_k), u_N(t_k))y_{N+M}(t_k) +$$
$$+ f_u(t_k, x^{u_N}(t_k), u_N(t_k))d_{N+M}(t_k)),$$
$$y_{N+M}(t_0) = 0, \; k = 0, 1, \dots, N + M - 1.$$

By $x^{u_N}(\cdot)$ we denote here the solution to the initial value problem (1.3) for $u_N(\cdot)$. The Sequential Proximal-Based Method (2.9) for (3.4) can be rewritten as follows ([20], p.12)

$$d_{N+M}(\cdot) = \mathcal{P}_{\tilde{J}, \mathcal{U}_{N+M} - u_N(\cdot), K_{N+M}}(\mathcal{P}^s_{\tilde{J}, \mathcal{U}_{N+M-1} - u_N(\cdot), K_{N+M-1}}(d_{N+M-1}(\cdot))),$$
$$s \in \mathbb{N}, \; \infty > \tilde{K} > K_{N+M} \downarrow K > 0, \; M = 1, 2, \dots, \quad (3.5)$$

where

$$\mathcal{P}_{\tilde{J}, \mathcal{U}_{N+M} - u_N(\cdot), K_{N+M}}(\mathcal{P}^s_{\tilde{J}, \mathcal{U}_{N+M-1} - u_N(\cdot), K_{N+M-1}}(d_{N+M-1}(\cdot))) =$$
$$= \text{Argmin}_{d(\cdot) \in \mathcal{U}_{N+M} - u_N(\cdot)} [\tilde{J}(d(\cdot)) + \frac{K_{N+M}}{2} \| d(\cdot) -$$
$$- \mathcal{P}^s_{\tilde{J}, \mathcal{U}_{N+M-1} - u_N(\cdot), K_{N+M-1}}(d_{N+M-1}(\cdot)) \|^2_{\mathbb{L}^{2,N+M}_m(G_{N+M})}].$$

The following theorem establishes the convergence properties of these finite-dimensional approximations ([20], Theorem 3, pp.13-14).

Theorem 11 *Let $\{d_{N+M}(\cdot)\}$ be the sequence generated by the method* (3.5). *Then*

$$\lim_{M \to \infty} \lim_{s \to \infty} \tilde{J}(d_{N+M}(\cdot)) = \tilde{J}(d^{opt}(\cdot))$$

and

$$\lim_{M \to \infty} \lim_{s \to \infty} \| d_{N+M}(\cdot) - \mathcal{P}^s_{\tilde{J}, \mathcal{U}_{N+M-1} - u_N(\cdot), K_{N+M-1}}(d_{N+M-1}(\cdot)) \|_{\mathbb{L}^2_m([0,1])} = 0.$$

If we use the Proximal-Penalty Method (2.10), we also deal with the optimization problem (3.4) and define the algorithm ([22], p.25)

$$d_{N+M}(\cdot) = \mathcal{P}_{\tilde{J}, \mathcal{U}_{N+M}-u_N(\cdot), K_{N+M-1}}(d_{N+M-1}(\cdot)) \qquad (3.6)$$

where

$$\mathcal{P}_{\tilde{J}, \mathcal{U}_{N+M}-u_N(\cdot), K_{N+M-1}}(d_{N+M-1}(\cdot)) := \mathrm{Argmin}_{d(\cdot)\in\mathcal{U}_{N+M}-u_N(\cdot)}[\tilde{J}(d(\cdot))+$$

$$+ \frac{K_{N+M-1}}{2}\|d(\cdot) - d_{N+M-1}(\cdot)\|^2_{\mathbb{L}^{2,N+M}_m(G_{N+M})}],$$

$$K_{N+M-1} > 0, \ \sum_{M=1}^{\infty}\frac{1}{\sqrt{K_{N+M-1}}} < \infty.$$

In [22] (Theorem 4.1. pp.26-27) we prove our next convergence result.

Theorem 12 *The sequence of problems* (3.4) *is an approximating sequence (in the sense of* Definition 3*) for the convex-linear problem* (3.2). *The sequence* $\{d_{N+M}(\cdot)\}$ *generated by the method* (3.6) *is a minimizing sequence for* (3.2). *This sequence converges strongly to an optimal solution of the optimal control problem* (3.2)

$$\lim_{M\to\infty}\|d_{N+M}(\cdot) - d^{opt}(\cdot)\|_{\mathbb{L}^2_m([0,1])} = 0.$$

Finally, note that the proofs of Theorem 11 and Theorem 12 use the properties of the proximal mapping (Section 2.1) and the discussed convex approximations (Section 2.3).

3.2 The General Gradient Formulaes

From the above discussion it follows that for constructive solving the given convex optimization problems (2.1) and (3.2) the corresponding proximal point method must be combined with some numerical procedures for the auxiliary convex problems (2.8) and (3.4). The development of the optimization theory has proceeded almost contemporarily with the systematical investigation of convex problems and their numerical treatment. A great amount of works is devoted to the theoretical and practical aspects of convex programming; see e.g., [118, 56, 199] and the references therein. It should be stressed that in the context of a regularized convex optimization problem one can consider a standard numerical algorithm. For purpose of numerical calculations, in [20, 22] we use a variant of the finite-dimensional gradient method. Evidently, a gradient-based computational approach can be simply implemented. Therefore, in this and the following sections we deal with the representation of the "reduced" gradient $\nabla \tilde{J}(u(\cdot))$ of the objective functional $\tilde{J}(u(\cdot))$ in (1.4) and with the auxiliary discrete OCP. Moreover, we consider the gradient $\nabla \mathcal{J}(u(\cdot))$ of the objective functional $\mathcal{J}(u(\cdot))$ in (1.4a). Let us first examine the abstract OCP (the generalization of (1.2)) which involves a control variable v along with the state variable ξ

$$\begin{aligned} &\text{minimize } T(\xi, v) \\ &\text{subject to } P(\xi, v) = 0, \\ &(\xi, v) \in S, \end{aligned} \tag{3.7}$$

where X, Y are real Banach and Hilbert spaces, $T : X \times Y \to R$ is an objective functional and

$$P : X \times Y \to X$$

is a given mapping. The terms mapping, function and operator will be used synonymously. Furthermore, derivatives considered here are Fréchet derivatives. By S we denote a nonempty subset of $X \times Y$. We say that an admissible pair

$$(\hat{\xi}, \hat{v}) \in Q := \{(\xi, v) \in S \mid P(\xi, v) = 0\}$$

is a local solution of (3.7) if

$$T(\hat{\xi}, \hat{v}) \leq T(\xi, v) \quad \forall (\xi, v) \in W_{(\hat{\xi}, \hat{v})} \subset Q,$$

where $W_{(\hat{\xi}, \hat{v})} \subset X \times Y$ is a neighborhood of $(\hat{\xi}, \hat{v})$. We assume that the mappings T and P are continuously differentiable and that the equation $P(\xi, v) = 0$ can be solved with respect to ξ, i.e. $\xi = \omega(v)$, where $\omega : Y \to X$ is a differentiable function. In this case the functional $T(\xi, v)$ can be represented as a functional depending only on v, namely,

$$T(\xi, v) = T(\omega(v), v) = \tilde{T}(v).$$

The introduced abstract optimal control problem (3.7) is of primary importance in many applications (see e.g., [120, 49, 124, 178]). Note that an ODE- or a PDE-optimal control problem or an optimal control problem with integral equations can also be formulated (in various ways) as an abstract problem (3.7) (see e.g., [194]). Moreover, a usual finite-dimensional approximation of an initial optimal control problem has a form of the minimization problem (3.7). In the above cases the condition $P(\xi, v) = 0$ represents the corresponding "state equation" of a specific optimal control problem.

Assume that the abstract problem (3.7) is *regular* (see [120]). Define the Lagrangian of problem (3.7)

$$\mathcal{L}(\xi, v, p) := T(\xi, v) + \langle p, P(\xi, v) \rangle_X,$$

where $p \in X^*$ and $\langle p, \cdot \rangle_X : X \to \mathbb{R}$. Here X^* is the (topological) dual space to X. For the Lagrange Multiplier Rule see, e.g., [120, 124] or [25]. We use the standard notation

$$T_\xi, \ T_v, \ P_\xi, P_v, \ \mathcal{L}_\xi, \ \mathcal{L}_p, \ \mathcal{L}_u$$

for the partial derivatives of the functions T, P and $\mathcal{L}$. Moreover, we introduce the adjoint operators

$$T_\xi^*, \ T_v^*, \ P_\xi^*, P_v^*, \ \mathcal{L}_\xi^*, \ \mathcal{L}_p^*, \ \mathcal{L}_u^*$$

to the corresponding linear operators and also consider the adjoint operator $\nabla \tilde{T}^*(v)$ to $\nabla \tilde{T}(v)$. Our next result ([26, ?], Theorem 7, pp.17-18) is an immediate consequence of the above solvability assumption for the state equation $P(\xi, v) = 0$. Note that a usual solvability criterion for this follows from appropriate variant of the Implicit Function Theorem [81]. For example, if the mapping P satisfies the condition of the Classic Implicit Function Theorem (see e.g., [191, 4]), then the equation $P(\xi, v) = 0$ is solvable. Some solvability conditions for a class of equations with *stable* differentiable operators P are studied in the book of the author [25].

Theorem 13 *Let T and P be continuously* Fréchet *differentiable and the equation in (3.7) is solvable. Assume that there exists the inverse operator*

$$(P_\xi^*)^{-1} \in L((X^* \times Y^*), X^*)$$

to P_ξ. Then the gradient $\nabla \tilde{T}^*(v)$ can be found by solving the following equations

$$P(\xi, v) = \mathcal{L}_p^*(\xi, v, p) = 0,$$
$$T_\xi^*(\xi, v) + P_\xi^*(\xi, v)p = \mathcal{L}_\xi^*(\xi, v, p) = 0, \tag{3.8}$$
$$\nabla \tilde{T}^*(v) = T_v^*(\xi, v) + P_v^*(\xi, v)p = \mathcal{L}_v^*(\xi, v, p).$$

Let us sketch the proof of Theorem 13. Differentiating the functional $\tilde{T}$ and state equation in (3.7) we obtain

$$P_\xi(\xi, v)\nabla\omega(v) + P_v(\xi, v) = 0,$$
$$\nabla\tilde{T}(v) = T_v(\xi, v) + T_\xi(\xi, v)\nabla\omega(v).$$

The existence of $(P_\xi^*)^{-1}$ implies the formula

$$\nabla\omega(v) = -(P_\xi)^{-1}(\xi, v)P_v(\xi, v).$$

Hence

$$\nabla\tilde{T}(v) = T_v(\xi, v) - T_\xi(\xi, v)(P_\xi)^{-1}(\xi, v)P_v(\xi, v),$$

and

$$\nabla\tilde{T}^*(v) = T_v^*(\xi, v) - P_v^*(\xi, v)(P_\xi^*)^{-1}(\xi, v)T_\xi^*(\xi, v). \tag{3.9}$$

On the other hand, we can calculate p from the second (adjoint) equation in (3.8) and substitute it to the third (gradient) equation. In this manner we also obtain the given relation (3.9). Note that a related result was also obtained by Bittner for a class of abstract optimal control problems in Banach spaces by using a technique of "directional limits" [50].

In the case of OCP (1.2) on $[0, 1]$ we have

$$X = \mathbb{W}_n^{1,1}([0, 1]), \ Y = \mathbb{L}_m^2([0, 1]).$$

By $\mathbb{W}_n^{1,1}([0, 1])$ we denote here the Sobolev space of absolutely continuous functions (see [12]). Note that instead of this Sobolev space one can consider $\mathbb{W}_n^{1,\infty}([0, 1])$, namely, the Sobolev space of differentiable functions with essentially bounded derivatives [12]. First we consider (1.2) without constraints. The objective functional T in (1.2) has the form $J(x(\cdot), u(\cdot))$, and the operator equation $P(\xi, v) = 0$ for (1.2) can be defined as follows

$$P(x(\cdot), u(\cdot)) := \left(\begin{array}{c} \dot{x}(t) - f(t, x(t), u(t)) \\ x(0) - x_0 \end{array} \right) = 0.$$

It is well known that in this particular case the evaluation of $\nabla\tilde{J}^*(u(\cdot))$ is relatively easy [178, 204]. For the unconstrained problem (1.2) with an integral objective functional we formulate the following theorem ([26], Theorem 8, p.18).

Theorem 14 *Under the condition of* Section 1.2 *the (reduced) gradient* $\nabla \tilde{J}^*(u(\cdot))$ *of the objective functional* $\tilde{J}(u(\cdot))$ *in a regular unconstrained problem* (1.4) *can be found by solving the equations*

$$\mathcal{L}_p^*(\xi, v, p) = 0 \Leftrightarrow \dot{x}(t) = H_p(t, x(t), u(t), p(t)), \ x(0) = x_0,$$
$$\mathcal{L}_\xi^*(\xi, v, p) = 0 \Leftrightarrow \dot{p}(t) = -H_x(t, x(t), u(t), p(t)), \ p(1) = 0,$$
$$\nabla \tilde{T}^*(v) = \mathcal{L}_v^*(\xi, v, p) \Leftrightarrow \nabla \tilde{J}^*(u(\cdot))(t) =$$
$$= -H_u(t, x(t), u(t), p(t)),$$

(3.10)

where $H(t, x, u, p) = \langle p, f(t, x, u) \rangle_{\mathbb{R}^n} - f_0(t, x, u)$ *is the* Hamiltonian *of the unconstrained problem* (1.2).

We now dwell briefly on the proof of Theorem 14 (see [26], pp.18-20). First let us recall that

$$\nabla \tilde{J}(u(\cdot)) \in L(\mathbb{L}_m^2([0, 1]), \mathbb{R}) = (\mathbb{L}_m^2([0, 1]))^* = \mathbb{L}_m^2([0, 1])$$

and $\nabla \tilde{J}^*(u(\cdot)) \in L(\mathbb{R}, \mathbb{L}_m^2([0, 1]))$. The Hilbert space $\mathbb{L}_m^2([0, 1])$ is identified with its dual. For every $g(\cdot) \in \mathbb{L}_m^2([0, 1])$ and every $\eta \in \mathbb{R}$ we have

$$\langle \nabla \tilde{J}(u(\cdot))g(\cdot), \eta \rangle_{\mathbb{R}} = \langle g(\cdot), \nabla \tilde{J}^*(u(\cdot))\eta \rangle_{\mathbb{L}_m^2([0,1])}$$

or

$$\eta \nabla \tilde{J}(u(\cdot))g(\cdot) = \int_0^1 \theta_\eta(t) g(t) dt,$$

where $\theta_\eta(\cdot) := \nabla \tilde{J}^*(u(\cdot))\eta \in \mathbb{L}_m^2([0, 1])$. For $\eta = 1$ we write

$$\nabla \tilde{J}(u(\cdot))g(\cdot) = \int_0^1 \theta(t) g(t) dt$$

This is consistent with the representation of $\nabla \tilde{J}(u(\cdot))$ as a function from $\mathbb{L}_m^2([0, 1])$. For some concrete calculations of the function θ see Example 4 and Example 5 from Section 3.4. The Lagrangian of the regular unconstrained problem (1.4) can be written as

$$\mathcal{L}(x(\cdot), u(\cdot), \hat{p}, p(\cdot)) = \int_0^1 f_0(t, x(t), u(t)) dt + \langle \hat{p}, x(0) - x_0 \rangle_{\mathbb{R}^n} +$$
$$+ \int_0^1 \langle p(t), \dot{x}(t) - f(t, x(t), u(t)) \rangle_{\mathbb{R}^n} dt$$

where the adjoint variable here contains two components $\hat{p}$ and $p(\cdot)$. For the unconstrained OCP (1.2) we introduce the usual Hamiltonian $H(t, x, u, p)$ (see Theorem 14). If we differentiate the Lagrange function with respect to the adjoint variable, then for every

$$g(\cdot) = (g_1, g_2(\cdot)), \ g_1 \in \mathbb{R}^n, \ g_2(\cdot) \in \mathbb{L}^2_m([0, 1])$$

we obtain

$$\mathcal{L}_{(\hat{p},p)}(x(\cdot), u(\cdot), \hat{p}, p(\cdot))g(\cdot) = \langle x(0) - x_0, g_1 \rangle_{\mathbb{R}^n} +$$

$$+ \int_0^1 \dot{x}(t) - f(t, x(t), u(t))g_2(t)dt = 0.$$

It means that

$$\mathcal{L}^*_{(\hat{p},p)}(x(\cdot), u(\cdot), \hat{p}, p(\cdot)) = (x(0) - x_0, \dot{x}(t) - f(t, x(t), u(t))).$$

and from the first equation of (3.8) we deduce that

$$\dot{x}(t) = f(t, x(t), u(t)) = H_p(t, x(t), u(t), p(t)), \ x(0) = x_0.$$

The first relation in (3.10) is proved. Consider the term

$$\int_0^1 \langle p(t), \dot{x}(t) \rangle_{\mathbb{R}^n} dt.$$

From the integration by part we have

$$\int_0^1 \langle p(t), \dot{x}(t) \rangle_{\mathbb{R}^n} dt = \langle p(1), x(1) \rangle_{\mathbb{R}^n} - \langle p(0), x(0) \rangle_{\mathbb{R}^n} -$$

$$- \int_0^1 \langle \dot{p}(t), x(t) \rangle_{\mathbb{R}^n}.$$

Hence

$$\mathcal{L}(x(\cdot), u(\cdot), \hat{p}, p(\cdot)) = \langle p(1), x(1) \rangle_{\mathbb{R}^n} + \langle \hat{p} -$$

$$- p(0), x(0) \rangle_{\mathbb{R}^n} - \langle \hat{p}, x_0 \rangle_{\mathbb{R}^n} - \int_0^1 \langle \dot{p}(t), x(t) \rangle_{\mathbb{R}^n} dt + \tag{3.11}$$

$$+ \int_0^1 f_0(t, x(t), u(t)) - \langle p(t), f(t, x(t), u(t)) \rangle_{\mathbb{R}^n} dt.$$

If we differentiate $\mathcal{L}$ in (3.11) with respect to $x(\cdot)$, we can compute $\mathcal{L}_x$, $\mathcal{L}_x^*$ and can deduce the second relation in (3.10). Using (3.11), we also write

$$\mathcal{L}_u(x(\cdot), u(\cdot), \hat{p}, p(\cdot))v(\cdot) = -\int_0^1 H_u(t, x(t), u(t))v(t)dt$$

for every $v(\cdot) \in \mathbb{L}_m^2([0, 1])$. Therefore, we obtain the last relation in (3.10)

$$\nabla \tilde{J}^*(u(\cdot))(t) = \mathcal{L}_u^*(x(\cdot), u(\cdot), \hat{p}, p(\cdot)) = -H_u(t, x(t), u(t)).$$

Recall that an OCP governed by ordinary differential equations can also be considered as a problem with a terminal objective functional

$$\mathcal{J}(u(\cdot)) := \phi(x^u(1)),$$

where ϕ is a differentiable function. The corresponding theorem for the *regular* OCP (1.2a) with a terminal objective functional $\mathcal{J}$ is proved in [178]. For some related results see also [204, 50]. We present here the computational formulae for the gradient $\nabla \mathcal{J}^*(u(\cdot))$ of the terminal objective functional $\mathcal{J}(u(\cdot))$ in the unconstrained OCP (1.2a) (see [178])

$$\dot{x}(t) = H_p(t, x(t), u(t), p(t)), \quad x(0) = x_0,$$
$$\dot{p}(t) = -H_x(t, x(t), u(t), p(t)), \quad p(1) = -\phi_x(x(1)), \qquad (3.12)$$
$$\nabla \mathcal{J}^*(u(\cdot))(t) = -H_u(t, x(t), u(t), p(t)).$$

Here
$$H(t, x, u, p) = \langle p, f(t, x, u) \rangle_{\mathbb{R}^n}$$

is the Hamiltonian of the unconstrained OCP (1.2a). In conclusion it may be said that the evaluation of the "reduced" Hessian for problem (1.4) or for problem (1.4a) is a fairly complex task. We refer to [178] for details.

3.3 Discretizations and Reduced Gradients

We now consider a discretization of the OCP (1.4) on the grid G_N and define

$$P(x(\cdot), u(\cdot)) := \left(\begin{array}{c} x_N(t_{k+1}) - \tilde{f}(t_k, x_N(t_k), u_N(t_k)) \\ x_N(t_0) - x_0 \end{array} \right), \quad k = 0, ..., N-1,$$

where $x_N(\cdot)$ is a solution of the discretized initial value problem (1.3) for

$$u_N(\cdot) \in \mathbb{L}_m^{2,N}(G_N).$$

The function $\tilde{f}$ is defined by the chosen discretization method for the initial differential equation. Moreover, the discretized objective functional is

$$\tilde{J}(u_N(\cdot)) := \sum_{k=0}^{N-1} f_0(t_k, x_N(t_k), u_N(t_k)) \Delta t_{k+1}$$

We assume that the initial problem (1.4) and all discrete problems are regular. Let us introduce the Lagrangian and the Hamiltonian for the discrete unconstrained OCP (1.4)

$$\mathcal{L}(x_N(\cdot), u_N(\cdot), \hat{p}, p_N(\cdot)) = \sum_{k=0}^{N-1} f_0(t_k, x_N(t_k), u_N(t_k)) \Delta t_{k+1} +$$

$$+ \langle \hat{p}, x_N(t_0) - x_0 \rangle_{\mathbb{R}^n} +$$

$$+ \sum_{k=0}^{N-1} \langle p_N(t_{k+1}), x_N(t_{k+1}) - \tilde{f}(t_k, x_N(t_k), u_N(t_k)) \rangle_{\mathbb{R}^n},$$

and

$$H(t_k, x_N(t_k), u_N(t_k), p_N(t_{k+1})) = \langle p_N(t_{k+1}), \tilde{f}(t_k, x_N(t_k), u_N(t_k)) \rangle_{\mathbb{R}^n} -$$
$$- f_0(t_k, x_N(t_k), u_N(t_k)),$$

where $p_N(\cdot)$ is the discrete adjoint variable. In order to get a final formula for the reduced gradient in the discrete OCP we rewrite the Lagrangian $\mathcal{L}$ in the form

$$\mathcal{L}(x_N(\cdot), u_N(\cdot), \hat{p}, p_N(\cdot)) = \langle \hat{p}, x_N(t_0) - x_0 \rangle_{\mathbb{R}^n} -$$

$$- \sum_{k=0}^{N-1} \left(H(t_k, x_N(t_k), u_N(t_k), p_N(t_{k+1})) - \langle p_N(t_{k+1}), x_N(t_{k+1}) \rangle_{\mathbb{R}^n} \right) \qquad (3.13)$$

Evidently, formula (3.13) is analogous to (3.11). This formula provides a basis for deducing the effective computational relations for gradient $\nabla \tilde{J}(u_N(\cdot))$ in the discretized OCP (1.2). If we differentiate $\mathcal{L}$ in (3.13) in much the same way as in Section 3.2, we can prove the following theorem (see also [26], Theorem 9, p.21).

Theorem 15 *Under the condition of* Section 1.2 *the (reduced) gradient* $\nabla \tilde{J}^*(u_N(\cdot))$ *of the objective functional* $\tilde{J}$ *in the discretized regular unconstrained* OCP (1.4) *can be found by solving the equations*

$$\mathcal{L}_p^*(\xi, v, p) = 0 \Leftrightarrow x_N(t_{k+1}) = H_p(t_k, x_N(t_k), u_N(t_k), p_N(t_{k+1})),$$

$$x_N(t_0) = x_0,$$

$$\mathcal{L}_\xi^*(\xi, v, p) = 0 \Leftrightarrow p_N(t_k) = H_x(t_k, x_N(t_k), u_N(t_k), p_N(t_{k+1})),$$

$$p_N(1) = 0,$$

$$\nabla \tilde{T}^*(v) = \mathcal{L}_v^*(\xi, v, p) \Leftrightarrow \nabla \tilde{J}^*(u_N(\cdot))(t_k) =$$

$$= -H_u(t_k, x_N(t_k), u_N(t_k), p_N(t_{k+1})),$$

(3.14)

where

$$H(t, x, u, p) = \langle p, \tilde{f}(t, x, u) \rangle_{\mathbb{R}^n} - f_0(t, x, u)$$

is the Hamiltonian *of the discretized unconstrained problem* (1.2).

We refer to [178] for computational formulae for the corresponding (reduced) gradient $\nabla \mathcal{J}(u_N(\cdot))$ of the discrete terminal objective functional

$$\mathcal{J}(u_N(\cdot)) := \phi(x^{u_N}(1))$$

from (1.4a).

Finally note that the possibility to obtain the relatively easy computational formulae for gradients given above follows from the representation of the objective functional as a function of the control variable only. A related evaluation of the full gradient for the functional $J(x(\cdot), u(\cdot))$ or its discrete variant is a fairly complex problem [195].

3.4 Algorithms and Examples

We continue by discussing some computational aspects of the proposed approach based on the proximal point algorithm and on the reduced gradient method. Our numerical techniques proposed in [20, 22] include at least two approximation steps, namely, the external linearization step and the internal discretization step. At first we choose an admissible initial control $u(\cdot)$ and linearize the initial control system (1.3) and the inequalities constraints in a neighborhood of the point $(x^u(\cdot), u(\cdot))$. We next fix the parameter $N \in \mathbb{N}$ and consider a family of the discrete optimization problems (3.4). Every problem (3.4) is a finite-dimensional convex-linear optimization problem such that the regularized functionals are not only convex but also *strongly convex* [127]. Note that the regularity conditions for the auxiliary problems (3.4) can be formulated as usual *Slater conditions*: there exists an inner point $d_{N+M}(\cdot)$ of the set

$$(\mathcal{U}_{N+M} - u_N(\cdot))$$

such that

$$\tilde{h}_j(u_N(\cdot)) + \langle \nabla \tilde{h}_j(u_N(\cdot)), d_{N+M}(\cdot) \rangle < 0 \ \forall j \in I,$$
$$\tilde{q}(u_N(\cdot))(t) + \langle \nabla \tilde{q}(u_N(\cdot))(t), d_{N+M}(\cdot) \rangle < 0 \ \forall t \in [0,1].$$

In the case of box-constraints (see Section 1.2)

$$U := \{u \in \mathbb{R}^m \ : \ b_-^i \le u_i \le b_+^i, \ i = 1, ..., m\},$$

we have $b_-^i - u_N(t_k) < d(t_k) < b_+^i - u_N(t_k)$ for all $t_k \in G_{N+M}$. Let

$$\mathcal{V}_{N+M} := \{v \in \mathcal{U}_{N+M} - u_N(\cdot) \ : \ \tilde{h}_j(u_N(\cdot)) + \langle \nabla \tilde{h}_j(u_N(\cdot)), v \rangle \le 0$$
$$\forall j \in I,$$
$$\tilde{q}(u_N(\cdot))(t) + \langle \nabla \tilde{q}(u_N(\cdot))(t), v \rangle \le 0 \ \forall t \in [0,1]\}$$

be the set of admissible controls for (3.4). If we use the Sequential Proximal-Based (3.5), we deal with the following auxiliary subproblems

$$\hat{J}(d(\cdot)) + \frac{K_{N+M}}{2} \|d(\cdot) - \rho_1\|_{\mathbb{L}_m^{2,N+M}(G_{N+M})}^2 \to \min, \tag{3.15}$$
$$d(\cdot) \in \mathcal{V}_{N+M},$$

and

$$\hat{J}(d(\cdot)) + \frac{K_{N+M-1}}{2} \|d(\cdot) - \rho_2\|_{\mathbb{L}_m^{2,N+M-1}(G_{N+M-1})}^2 \to \min, \tag{3.16}$$
$$d(\cdot) \in \mathcal{V}_{N+M-1},$$

where

$$\rho_1 := \mathcal{P}_{\hat{J}, \mathcal{U}_{N+M-1} - u_N(\cdot), K_{N+M-1}}^s (d_{N+M-1}(\cdot)),$$
$$\rho_2 := \mathcal{P}_{\hat{J}, \mathcal{U}_{N+M-1} - u_N(\cdot), K_{N+M-1}}^\tau (d_{N+M-1}(\cdot))$$

for $\tau < s$. In the case of the Proximal-Penalty Method (3.6) we have the auxiliary subproblem

$$\hat{J}(d(\cdot)) + \frac{K_{N+M-1}}{2}\|d(\cdot) - d_{N+M-1}(\cdot)\|^2_{\mathbb{L}^{2,N+M}_m(G_{N+M})} \to \min,$$

$$d(\cdot) \in \mathcal{V}_{N+M}. \tag{3.17}$$

Note that in the finite-dimensional cases (3.15)-(3.17) the minimizing sequences generated by the classic proximal point method convergence strongly to an optimal solution of the initial problem. There is no need to use a strong convergent modification of the classic proximal method from Section 2.2. In [20, 22] we solve the formulated auxiliary problems (3.15)-(3.17) by applying the Projected Gradient Algorithm (see e.g., [100, 172, 173])

$$d^{l+1}(t_k) = P_\mathcal{V}(d^l(t_k) - \gamma_l(\nabla \hat{J}(d^l(\cdot))(t_k) + F(t_k))),$$

$$l = 0, 1, \dots \tag{3.18}$$

where γ_l is a *step-size* and $P_\mathcal{V}$ is a projection on the set $\mathcal{V}$. We have $\mathcal{V} \equiv \mathcal{V}_{N+M}$ for (3.15), (3.17) and $\mathcal{V} \equiv \mathcal{V}_{N+M-1}$ for (3.16). Moreover,

$$F(\cdot) \equiv K_{N+M}(d^l(\cdot) - \rho_1)$$

for (3.15),

$$F(\cdot) \equiv K_{N+M-1}(d^l(\cdot) - \rho_2)$$

for (3.16) and

$$F(\cdot) \equiv K_{N+M-1}(d^l(\cdot) - d_{N+M-1}(\cdot))$$

in the case of (3.17). The gradient $\nabla \hat{J}(d^l(\cdot))$ can be computed with methods presented in Section 3.2 and Section 3.3. For example, this gradient in (3.15) is as follows

$$\nabla \hat{J}(d^l(\cdot))(t_k) = -H_d(t_k, y_{N+M}(t_k), u_N(t_k), d^l(t_k), p(t_{k+1})),$$

where $t_k \in G_{N+M}$, $k = 1, \dots N + M - 1$, $l = 0, 1, \dots$ and

$$p(t_k) = H_y(t_k, y_{N+M}(t_k), u_N(t), d^l(t_k), p(t)), \quad p(1) = 0.$$

The Hamiltonian of the discrete unconstrained problem (3.4) can be written in the following way

$$H(t_k, y_{N+M}(t_k), u_N(t_k), d^l(t_k), p(t_{k+1})) :=$$

$$= -f_0(t_k, y_{N+M}(t_k), u_N(t_k) + d^l(t_k))\Delta t_{k+1}+$$

$$+ \langle p(t_{k+1}), \Delta t_{k+1}(f_x(t_k, x^{u_N}(t_k), u_N(t_k))y_{N+M}(t_k)+$$

$$+ f_u(t_k, x^{u_N}(t_k), u_N(t_k))d^l(t_k) + y_{N+M}(t_k)\rangle_{\mathbb{R}^n},$$

$$k = 0, \dots, N + M - 1.$$

We refer to [100, 172, 173] for the convergence properties of (3.18). For choosing the step-size γ_l consult [100, 173, 101]. Compare to the Sequential Proximal-Based (3.5), the Proximal-Penalty Method (3.6) is sophisticated procedure. On the other hand, in (3.6) we have

$$\frac{1}{\sqrt{K_{N+M-1}}} \to 0. \tag{3.19}$$

It is well known that (3.19) is an adverse condition from the point of view of concrete computations. For a fixed $N \in \mathbb{N}$ and $u_N(\cdot)$ we can consider a sequence generated by the Sequential Proximal-Based (3.5) or by the Proximal-Penalty Method (3.6). Let

$$d_{N+M}(\cdot) \in \mathcal{U}_{N+M} - u_N(\cdot)$$

be an element of this sequence such that

$$|\hat{J}(d_{N+M}(\cdot)) - \hat{J}(d_{N+M-1}(\cdot))| < \delta_{M,N},$$

where $\delta_{M,N} > 0$ is a sufficiently small real number. We now put

$$u_{N+M}(\cdot) := u_N(\cdot) + d_{N+M}(\cdot).$$

It is evident that $u_{N+M}(\cdot) \in \mathcal{U}_{N+M}$. Given this control function the solution of the initial value problem (1.3) is denoted by $x^{u_{N+M}}(\cdot)$. Using $u_{N+M}(\cdot)$ and the computed trajectory $x^{u_{N+M}}(\cdot)$, we can consider the next linearization step. Note that the theoretical results presented in Section 3.1 guarantee only the good convergence properties of the proposed discrete approximations. The global convergence of the general linear-discrete numerical scheme critically depends on the considered linearization step and on the corresponding starting point.

Let us now illustrate the application of the proximal-type algorithms given above to two ill posed OCPs (see [20, 22])

Example 4 *Consider*

$$\text{minimize } J(u(\cdot)) := \int_0^1 x^2(t)dt$$

subject to $\dot{x}(t) = u(t)$, a.e. on $[0, 1]$,

$x(0) = 0$,

$u(\cdot) \in \mathbb{L}^2([0, 1])$, $|u(t)| \leq 1$,

$x(1) \leq 0$.

This OCP has a unique optimal solution $u^{opt}(t) = 0$ a.e., the objective functional

$$J(u(\cdot)) = \int_0^1 \left(\int_0^t u(\tau)d\tau \right)^2 dt$$

is convex and the set of admissible control functions is bounded, convex and closed. We apply the Sequential Proximal-Based algorithm (3.5) for

$$N = 50, \quad 1 \le M \le 50.$$

For $N = 50$, $M = 50$ the computed optimal control $\{u_{N+M}(\cdot)\}$ has the following property $\max_{0 \le k \le N+M} |u_{N+M}(t_k)| = 0.0016521$. The constraints were satisfied with tolerance 10^{-4}. The computed optimal objective value is $8 \cdot 10^{-7}$. Note that the Hamiltonian of the unconstrained problem is $H(x, u, p) = pu - x^2$. We have $\dot{p}(t) = 2x(t)$, $p(1) = 0$, and

$$p(t) = 2\left(- \int_0^1 x(t)dt + \int_0^t x(\tau)d\tau\right).$$

Hence

$$\nabla J(u(\cdot))(t) = -H_u(x(t), u(t), p(t)) = -p(t) =$$

$$= 2\left(\int_0^1 x(t)dt - \int_0^t x(\tau)d\tau\right).$$

The optimal solution $u^{opt}(\cdot)$ satisfies the condition

$$\nabla J(u^{opt})(t) = 0.$$

Example 5 *We now deal with the OCP of the type* (1.2a)

$$\text{minimize } \mathcal{J}(u(\cdot)) = x^2(1)$$
$$\text{subject to } \dot{x}(t) = u(t), t \in (0, 1],$$
$$x(0) = 0,$$
$$u(\cdot) \in \mathbb{L}^2([0, 1]), |u(t)| \le 1,$$
$$x(1) \le -1/2.$$

This optimal control problem also has a unique solution $u^{opt}(t) = -1/2$. Clearly, the objective functional $\mathcal{J}(u(\cdot))$ is convex and the set of admissible control functions is bounded, convex and closed. We use the Proximal-Penalty Method (3.6) with the following parameters

$$K_{N+M} = (N + M)^2, \quad N = 50, \quad 1 \le M \le 50.$$

For $M = 50$ the constraints were satisfied with the tolerance 10^{-5}. For the computed optimal controls $\{u_{N+M}(\cdot)\}$ we have the estimate

$$\max_{0 \le k \le N+M-1} |u_{N+M}(t_k) - u^{opt}(t_k)| = 0.0271.$$

The computed optimal objective values is 0.27074. In this case we have

$$H(x, u, p) = pu,$$
$$\dot{p}(t) = 0,$$
$$p(1) = -2x(1).$$

Hence $p(t) = -2x(1)$. The gradient $\nabla\mathcal{J}(u(\cdot))$ can be written as follows

$$\nabla\mathcal{J}(u(\cdot)) = -p = 2x(1).$$

Evidently, the direct computation of the gradient of $\mathcal{J}(u(\cdot))$ involves the same result.

In both examples the basis for the concrete computational method is the classic gradient method applied to the regularized objective functional. We use the constant step-size $\gamma = 0.2$. The implementation of the computational schemes, described above, was carried out, using the "Numerical Recipes in C" package [177] and the author's program written in C.

3.5   Concluding Remarks

The introduced proximal-like method can be combined not only with the used gradient-type method but also with a second order optimization method (see [173]). An interesting variant of the Projected Gradient Algorithm for a class of convex-linear problems is presented in [187]. We have considered the optimal control problem under convexity assumptions. Using a variant of the proximal point method for nonconvex optimization [127], one can extend the presented proximal-type methods to some classes of nonconvex optimal control problems of the type (1.2). Finally note, that it seems to be possible to apply the proposed numerical technique to some classes of OCPs governed by partial differential equations.

Chapter 4

Convex Control Systems

We introduce so-called *convex control systems* and study the corresponding *convex optimal control problems*. The proximal-based algorithms for these convex OCPs can be analyzed in the same way as algorithms presented in Chapter 3.

4.1 Semilinear Problems

As a preliminary we present some easy concepts and facts presented, for example, in [183] and in [26] (Lemma 2, Lemma 3, p.6 and p.11).

Definition 4 *We say that a functional $\varpi : \Gamma \subset \mathbb{R}^n \to \mathbb{R}$ is monotonically nondecreasing if $\varpi(\xi) \geq \varpi(\zeta)$ for all $\xi, \zeta \in \Gamma$ such that*

$$\xi_k \geq \zeta_k, \ k = 1, ..., n.$$

Clearly, for $n = 1$ we obtain the usual monotonicity concept. It is evident that a linear positive functional is monotonically nondecreasing in the above sense. Let

$$\varpi(\xi) = \xi_1^2 + ... + \xi_n^2$$

with $\xi_k \in \mathbb{R}_+$. From $\xi_k \geq \zeta_k \geq 0$ for all $k = 1, ..., n$ it follows that $\varpi(\xi) \geq \varpi(\zeta)$.

Lemma 1 *Let $\varpi^1 : \mathbb{R}^n \to \mathbb{R}$ be a monotonically nondecreasing, convex functional. Assume that for every $k = 1, ..., n$ functions*

$$\varpi_k^2 : \mathbb{R}^m \to \mathbb{R}$$

are convex. Then the functional

$$\varpi : \mathbb{R}^m \to \mathbb{R}, \ \varpi(\cdot) := \varpi^1(\varpi^2(\cdot)),$$

where $\varpi^2(\xi) := (\varpi_1^2(\xi), ..., \varpi_n^2(\xi))^T$, $\xi \in \mathbb{R}^m$, is convex.

Lemma 2 *Let $\varpi^1 : \mathbb{R}^n \to \mathbb{R}$ be a convex functional and*

$$\varpi^2 : H \to \mathbb{R}^n$$

be a linear function on a Hilbert *space H. Then the functional*

$$\varpi : H \to \mathbb{R}, \ \varpi(\cdot) := \varpi^1(\varpi^2(\cdot))$$

is convex.

Definition 5 *We call the control system* (1.3) *a convex control system if every functional*

$$V_k(u(\cdot)) := x_k^u(t), \ u(\cdot) \in \mathcal{U}, \ t \in [0, t_f], \ k = 1, ..., n$$

is convex.

We are interested in studying the convex control systems in the context of optimal control problems. Therefore, we extend Definition 5 (see [26, 30]).

Definition 6 *If the infinite-dimensional minimization problem* (1.4) *(or* (1.4a)) *is equivalent to a convex optimization problem* (2.1), *then we call* (1.2) *(or* (1.2a)) *a convex optimal control problem.*

Let us now examine the initial optimal control problem (1.2a) with the terminal functional $\mathcal{J}(u(\cdot)) := \phi(x^u(1))$ instead of the integral functional J and with the semilinear (affine) differential equations (1.3), namely, with the right-hand side

$$f(t, x, u) = A(t)x + B(t, u).$$

We assume that $\phi(\cdot)$ is a differentiable function, $A(t) = (a_{i,j}(t))_n^n$ are regular $n \times n$ matrices for every $t \in [0, 1]$ and the functions $a_{i,l}(\cdot), \ i, l = 1, ..., n$ are continuous. The ensuing analysis is restricted to a continuous function $B : \mathbb{R} \times \mathbb{R}^m \to \mathbb{R}^n$. Note that under the given assumptions the initial value problem (1.3) has a unique solution $x^u(\cdot)$. In parallel with this OCP we also introduce the corresponding optimization problem (1.4a). A control system (1.3) governed by semilinear differential equations is called a *semilinear control system*, we refer to [108] for some theoretical facts about abstract semilinear control systems in Banach spaces. The above Lemma 1 makes it possible to formulate the following theorem (see [26], Theorem 1, pp.7-8).

Theorem 16 *Let $a_{i,j}(t) \geq 0$ for all $t \in [0,1]$ and all $i, j = 1, ..., n$. Assume that the functionals $u(\cdot) \rightarrow B_k(t, u(t))$, $u(\cdot) \in \mathcal{U}$ are convex for all indexes $k = 1, ..., n$ and all $t \in [0, t_f]$. Then the corresponding semilinear control system (1.3) is convex. If in addition to this the functionals $\phi(\cdot)$, $h(\cdot)$ are convex and monotonically nondecreasing and the functional $q(t, \cdot)$ is convex and monotonically nondecreasing for every $t \in [0, t_f]$, then (1.2a) is a convex OCP.*

Note that a linear or linearized control system also provides an example of a convex control system. It is well known, that linear control systems and the corresponding problems (1.2) have been subjects of the *robust control theory* (see e.g, [7] and [28]). In this connection one can introduce the convex minimax optimal control problem mentioned in [26, 30]. Using Lemma 2, we can prove a result for linear control systems (see [26], Theorem 5, pp.14-15).

Theorem 17 *Let the control system in (1.2a) be linear, i.e.*

$$f(t, x, u) = A(t)x + B(t)u,$$

where $A(t) \in \mathbb{R}^{n \times n}$, $B(t) \in \mathbb{R}^{n \times m}$ are regular matrices. If the functionals $\phi(\cdot)$, $h(\cdot)$ are convex and the functional $q(t, \cdot)$ is convex for all $t \in [0, t_f]$, then problem (1.4a) is a convex OCP.

Evidently, in the cases of Theorem 16 and Theorem 17 we have

$$Z = \mathbb{L}_m^2([0, t_f]),$$

here Z is a Hilbert space for problem (2.1).

4.2 Convex Control Systems

In this section we give some examples of these systems and discuss the associated convex OCP from [26, 30]. First let us formulate the following result [26, 30].

Theorem 18 *Assume that the function $f(t, x, u)$ in (1.3) is continuous and satisfies the following* Lipschitz *condition (uniformly in $u \in U$)*

$$\|f(t, x_1, u) - f(t, x_2, u)\| \leq L\|x_1 - x_2\|, \quad \forall x_1, x_2 \in \mathbb{R}^n, \ u \in U,$$

where $t \in [0, t_f]$. Let $f_k(t, \omega)$, $k = 1, ..., n$, be convex functionals with respect to the variable $\omega := (x, u)$ for every $t \in [0, t_f]$. Moreover, let $f_k(t, \cdot, u)$, $k = 1, ..., n$ be monotonically nondecreasing functionals for every $t \in [0, t_f]$, $u \in U$. Then the control system (1.3) is convex.

The next result is an immediate consequence of Theorem 18 (see [30]).

Theorem 19 *Consider the control system (1.3) with*

$$f(t, x, u) = f^1(t, x) + B(t)u,$$

where $f^1(t, \cdot)$ is a Lipschitz*-continuous function for all $t \in [0, t_f]$. Let $f_k^1(t, \cdot)$, $k = 1, ..., n$ be a convex and monotonically nondecreasing function for every $t \in [0, t_f]$. Assume that $u(t)$ is an m-dimensional control vector and $B(t) = (b_{i,j}(t))_m^n$ is an $n \times m$ matrix for every $t \in [0, t_f]$ such that all functions $b_{ij}(\cdot)$ are continuous. Then (1.3) is convex.*

It is easy to see that the function f satisfies all assumptions of Theorem 18. Hence, the corresponding control system (1.3) is convex. Note that a dynamical system with the above right-hand side can also be considered as a partial linearization (with respect to the control variable u) of a given nonlinear control system. Theorem 19 also follows from the following result [30].

Theorem 20 *Consider the control system (1.3) with*

$$f(t, x, u) = f^1(t, x) + f^2(t, u),$$

where $f^1(t, \cdot)$ is Lipschitz*-continuous function for all $t \in [0, t_f]$, i.e.*

$$\|f^1(t, x_1) - f^1(t, x_2)\| \leq L\|x_1 - x_2\|, \quad \forall x_1, x_2 \in \mathbb{R}^n.$$

Let $f_k^1(t, \cdot)$, $f_k^2(t, \cdot)$ $k = 1, ..., n$ be convex and $f_k^1(t, \cdot)$ $k = 1, ..., n$ be monotonically nondecreasing functionals for every $t \in [0, t_f]$. Then (1.3) is convex.

Example 6 *By way of a simple example let us consider the following one-dimensional control system*

$$\dot{x}(t) = x(t)u(t), \ x(0) = 1.$$

on the time-interval $[0, 1]$ *and the corresponding trajectories*

$$x^u(t) = \exp\left(\int_0^t u(\tau)d\tau\right), \ t \in [0, 1].$$

We deal here with the convex function $\exp(\cdot)$ *and with the linear integral-mapping.* Lemma 2 *implies the convexity of the functionals*

$$V(u(\cdot)) := x^u(t)$$

for all admissible control functions $u(\cdot) \in \mathcal{U}$ *and for all* $t \in [0, t_f]$.

It is well known that some classical nonlinear differential equations can be reduced, by a substitution, to a linear equation. If we deal with the controllable Bernoulli differential equation

$$\dot{x}(t) + w(t)x(t) + u(t)\sqrt{x(t)} = 0, \ x(0) = x_0, \tag{4.1}$$

then one can apply the substitution $\xi := \sqrt{x}$ and rewrite (4.1) into

$$\dot{\xi}(t) + 0.5w(t)\xi(t) + 0.5u(t) = 0, \ \xi(0) = \sqrt{x_0},$$

and the corresponding solution $\xi^u(t)$ is convex with respect to the control $u(\cdot) \in \mathcal{U}$ for all $t \in [0, t_f]$. Using Lemma 2, we obtain the convexity of the solution

$$x^u(t) = (\xi^u(t))^2$$

to the initial Bernoulli equation (4.1). Finally, note that the Bernoulli differential equation enjoys wide application in the *population dynamic* (see e.g., [17] and references therein).

Example 7 *Let us discuss the mathematical model of prolonged fishing (communicated by* Kugelmann)

$$\dot{x}_1(t) = w_1 x_1(t)\left(1 - \frac{x_1(t)}{K_1}\right) - \beta_{12}\frac{x_1(t)x_2(t)}{K_1 K_2} - q_1 x_1(t)u_1(t),$$

$$x(0) = x_{10} > 0,$$

$$\dot{x}_2(t) = w_2 x_2(t)\left(1 - \frac{x_2(t)}{K_2}\right) - \beta_{21}\frac{x_2(t)x_1(t)}{K_1 K_2} - q_2 x_2(t)u_2(t),$$

$$x_2(0) = x_{20} > 0,$$

$$\tag{4.2}$$

where $u(t) \in [0, 1]$ $\forall t \in [0, 1]$ and $w_1, w_2, K_1, K_2, q_1, q_2$ are positive constants. We divide the first equation by $(-x_1(t))$, the second equation by $(-x_2(t))$ and use the substitution $y_1 := \ln x_1$, $y_2 := \ln x_2$. Thus (4.2) can be rewritten as follows

$$
\begin{aligned}
\dot{y}_1(t) &= w_1\left(\frac{\exp(y_1(t))}{K_1} - 1\right) - \beta_{12}\frac{\exp(y_2(t))}{K_1 K_2} + q_1 u_1(t), \\
y(0) &= \exp x_{10}, \\
\dot{y}_2(t) &= w_2\left(\frac{\exp(y_2(t))}{K_2} - 1\right) - \beta_{21}\frac{\exp(y_1(t))}{K_1 K_2} + q_2 u_2(t), \\
y_2(0) &= \exp x_{20}.
\end{aligned}
\tag{4.3}
$$

By Theorem 20, the control system (4.3) is convex.

The convex control systems involve the convex OCPs. Our next result (see [26, 30]) establishes the convexity of an OCP (1.2a) with a terminal functional $\mathcal{J}$.

Theorem 21 Let the control system in (1.3) be convex and the functionals $\phi(\cdot)$, $h(\cdot)$ be convex and monotonically nondecreasing. Let the function $q(t, \cdot)$ be convex and monotonically nondecreasing for every $t \in [0, t_f]$. Then the corresponding OCP (1.2a) is convex.

Example 8 Consider a semilinear (affine) control system from Theorem 16 and formulate the following constrained OCP

$$
\begin{aligned}
&\text{minimize } vx(t_f) + \frac{1}{2}\int_0^{t_f} u^T(t)S u(t)dt \\
&\text{subject to } \dot{x}(t) = A(t)x + B(t, u), \\
&x(0) = x_0, \ u(t) \in U \ t \in [0, t_f], \\
&c(t)x(t) \leq d(t) \ \forall t \in [0, t_f],
\end{aligned}
\tag{4.4}
$$

where $v \in \mathbb{R}^n$, $v_i \geq 0$ for all $i = 1, ..., n$. Here S is an $m \times m$ symmetric positive definite weighting matrix and

$$
c(t) \in \mathbb{R}^n, \ d(t) \in \mathbb{R}
$$

for all times $t \in [0, t_f]$. Let $c_i(t) \geq 0$ for all

$$t \in [0, t_f], \ i = 1, ..., n.$$

Assume that there exists a control function $u(\cdot) \in \mathcal{U}$ such that the corresponding state $x^u(\cdot)$ satisfies the given linear inequality constraints. Introducing an additional state x_{n+1} given by the following initial value problem

$$\dot{x}_{n+1}(t) = \frac{1}{2} u^T(t) S u(t), \ t \in [0, t_f],$$

$$x_{n+1}(0) = 0,$$

(4.5)

we reduce (4.4) to an equivalent OCP with

$$\phi(\tilde{x}(t_f)) = v x(t_f) + x_{n+1}(t_f),$$

where

$$\tilde{x} := (x, x_{n+1})^T$$

is an extended state vector. Evidently, the right-hand side of the differential equations in (4.4)-(4.5) satisfies the convexity condition from Theorem 20 and the extended control system is also convex. Since the introduced linear cost functional $\phi(\cdot)$ is monotonically nondecreasing and the linear state constraints satisfy the corresponding conditions of Theorem 21, the given problem (4.4) is a convex OCP.

From the view-point of numerical mathematics we solve an OCP (1.2) or the equivalent optimization problem (1.4) approximately. A suitable numerical method for (1.4) generates a minimizing sequence so that the obtained approximations are consistent. Therefore, instead of the exact "equivalence" in the sense of Definition 6, one can restrict the consideration to a reasonable approximate concept. We now introduce a generalization of Definition 6.

Definition 7 *If for (1.2) (for (1.2a)) there exists a sequence of convex optimization problems (2.1) such that the solutions to (2.1) give rise to a minimizing sequence for the corresponding problem (1.4) (or (1.4a)), then (1.2) (or (1.2a)) is called an approximately convex OCP.*

We use Definition 7 and give some examples of the approximately convex OCP in the next chapter.

4.3   The Proximal-Based Algorithms

Consider a convex OCP of the type (1.2).   If we apply the classical proximal point method to (1.2), we deal with the following subproblem

$$\text{minimize } \tilde{J}(u(\cdot)) + \frac{\chi_r}{2}\|u(\cdot) - u^r(\cdot)\|^2_{\mathbb{L}^2_m([0,t_f])} =$$

$$= \int_0^{t_f} f_0(t, x^u(t), u(t))dt + \frac{\chi_r}{2}\int_0^1 \|u(t) - u^r(t)\|^2_{\mathbb{R}^m}dt$$

subject to $\dot{x}(t) = f(t, x(t), u(t))$ a.e. on $[0, t_f]$, (4.6)

$x(0) = x_0$,

$u(t) \in U$ a.e. on $[0, t_f]$,

$h_j(x(t_f)) \leq 0 \ \forall j \in I$,

$q(t, x(t)) \leq 0 \ \forall t \in [0, t_f]$,

where $r = 0, 1, \ldots$ and $u^0(\cdot) \in \mathcal{U}$ is an admissible control. Since (4.6) is a convex minimization problem with a strongly convex objective functional, we have a unique global minimizer. Using the strong convergent proximal-like methods presented in Section 2.2, one can use the solution of problem (4.6) and construct a strong convergent minimizing sequence. The auxiliary problem (4.6) can be solved directly by applying a standard minimization algorithm (see e.g., [100, 101, 172, 173, 199, 118, 15]). For example, we consider an infinite-dimensional version of the gradient method

$$u^{l+1}(t) = P_{\mathcal{V}}(u^l(t) - \gamma_l(\nabla \tilde{J}(u^l(\cdot)))(t) + \chi_l(u^l(t) - u^r(t)),$$

$$l = 0, 1, \ldots \tag{4.7}$$

where γ_l is a step-size and $P_{\mathcal{V}}$ is a projection on the set

$$\mathcal{V} := \{v(\cdot) \in \mathcal{U} \ : \ \tilde{h}_j(v) \leq 0 \ \forall j \in I, \ \tilde{q}(v)(t) \leq 0 \ \forall t \in [0, t_f]\}.$$

The gradient $\nabla \tilde{J}(u^l(\cdot))$ of the unconstrained problem (4.6) can be found from (3.10) (Section 3.2 Theorem 14). We refer to [117] for Conjugate Direction Methods, to [111] for the Gradient Plus Projection Method and to [178] for Feasible Directions Algorithms.

Discrete approximations are commonly encountered procedures in optimal control. A proximal-based approach can also be applied to a discretization of the convex OCP (1.2)

$$
\begin{aligned}
&\text{minimize } \tilde{J}(u_N(\cdot)) \\
&\text{subject to } u_N(\cdot) \in \mathcal{U}_N, \\
&\tilde{h}_j(u_N(\cdot)) \le 0 \ \forall j \in I, \\
&\tilde{q}(u_N(\cdot))(t) \le 0 \ \forall t \in [0,1].
\end{aligned}
\tag{4.8}
$$

Evidently, (4.8) is analogous to problem (3.3). However, the linearization step proposed in Chapter 3 does not need to be used for (4.8)! Let us present the Sequential Proximal-Based (2.9) for the convex problem (4.8)

$$
\tilde{J}(u(\cdot)) + \frac{K_N}{2} \|u(\cdot) - \rho_1\|^2_{\mathbb{L}^{2,N}_m(G_N)} \to \min,
$$
$$
u(\cdot) \in \mathcal{V}_N,
$$

and

$$
\tilde{J}(u(\cdot)) + \frac{K_{N-1}}{2} \|u(\cdot) - \rho_2\|^2_{\mathbb{L}^{2,N-1}_m(G_{N-1})} \to \min,
$$
$$
u(\cdot) \in \mathcal{V}_{N-1},
$$

where

$$
\rho_1 := \mathcal{P}^s_{\tilde{J}, \mathcal{U}_{N-1}, K_{N-1}}(u_{N-1}(\cdot)), \ \rho_2 := \mathcal{P}^\tau_{\tilde{J}, \mathcal{U}_{N-1}, K_{N-1}}(u_{N-1}(\cdot)),
$$
$$
\mathcal{V}_N := \{v \in \mathcal{U}_N : \tilde{h}_j(v) \le 0 \ \forall j \in I, \ \tilde{q}(v)(t) \le 0 \ \forall t \in [0,1]\}.
$$

for $\tau < s$. If we apply the Proximal-Penalty Method (2.10) to the convex problem (4.8), we have

$$
\tilde{J}(u(\cdot)) + \frac{K_{N-1}}{2} \|u(\cdot) - u_{N-1}(\cdot)\|^2_{\mathbb{L}^{2,N}_m(G_N)} \to \min,
$$
$$
u(\cdot) \in \mathcal{V}_N.
$$

Since the proofs of Theorem 11 and Theorem 12 use only the convex structure of the convex-linear problem (3.4) (see also [20], Theorem 3, pp.13-14 and [22], Theorem 4.1. pp.26-27), we are in a position to establish the convergence results for the proximal-based methods presented above.

Theorem 22 *Let $\{u_N(\cdot)\}$ be the sequence generated by the Sequential Proximal-Based (2.9) for (4.8). Then*

$$\lim_{N\to\infty} \lim_{s\to\infty} \tilde{J}(u_N(\cdot)) = \tilde{J}(u^{opt}(\cdot))$$

and

$$\lim_{N\to\infty} \lim_{s\to\infty} \|u_N(\cdot) - \mathcal{P}^s_{\tilde{J},\mathcal{U}_{N-1},K_{N-1}}(u_{N-1}(\cdot))\|_{\mathbb{L}^2_m([0,1])} = 0.$$

Theorem 23 *The sequence of problems (4.8) is an approximating sequence (in the sense of Definition 3) for the convex OCP (4.6). The sequence $\{u_N(\cdot)\}$ generated by the Proximal-Penalty Method (2.10) for problem (4.8) is a minimizing sequence for (4.6). This sequence converges strongly to an optimal solution of the optimal control problem (4.6)*

$$\lim_{N\to\infty} \|u_N(\cdot) - u^{opt}(\cdot)\|_{\mathbb{L}^2_m([0,1])} = 0.$$

Note that for computing the reduced gradient $\nabla \tilde{J}(u(\cdot))$ of the unconstrained problem (4.6) we use the Hamiltonian

$$H^r(t,x,u,p) = \langle p, f(t,x,u)\rangle_{\mathbb{R}^n} - \left(f_0(t,x,u) + \frac{\chi}{2}\|u - u^r\|^2_{\mathbb{R}^m}\right)$$

of the unconstrained problem (4.6). This Hamiltonian coincides with the *augmented Hamiltonian* of the well-known Sakawa algorithm for OCPs [192, 55]. It seems plausible that in the case of the convex OCPs one can generalize the convergence result of Bonnans [55] for Sakawa-type algorithms and obtain a global convergence result.

It is well known that necessary optimality conditions for the above convex minimization problems are also sufficient. Moreover, a local solution of a convex minimization problem coincides with a global solution. Consider a convex OCP and the associated minimization problem. In the case of a convex OCP one also can formulate necessary optimality conditions in the form of the usual Pontryagin maximum principle (see e.g., [54, 44, 120, 94]).

When solving OCPs with state constraints based on some necessary conditions for optimality one is often faced with two technical difficulties: the irregularity of the Lagrange multiplier associated with the state constraint [80] and the *degeneracy phenomenon* (see e.g., [10]). Various supplementary conditions (constraint qualifications) have been proposed under which it is possible to assert that the Lagrange multiplier rule holds in "normal" form, i.e. that the corresponding minimization problem is "regular". We refer to [10, 96, 180] for details. In the following, we assume that problem (1.2a) is regular.

Theorem 24 *Consider a regular convex* OCP *(1.2a). An admissible control function* $u^{opt}(\cdot)$ *is an optimal solution of the convex problem* *(1.2a) if and only if there is a* Lagrange *multiplier* $\mu \geq 0$ *such that*

$$\mathcal{L}(u^{opt}(\cdot), \mu) = \min_{u(\cdot) \in Q_2 \cap \mathcal{U}} \mathcal{L}(u(\cdot), \mu),$$

$$\mu \tilde{h}(u^{\hat{o}pt}(\cdot)) = 0,$$

where $\mathcal{L}(u(\cdot), \mu) = \tilde{J}(u(\cdot)) + \mu \tilde{h}(u(\cdot))$ *is the Lagrange function and*

$$Q_2 := \bigcap_{t \in [0, t_f]} Q_2(t),$$

$$Q_2(t) := \{u(\cdot) \in \mathbb{L}_m^2([0, t_f]) \ : \ \tilde{q}(u(\cdot))(t) \leq 0 \ t \in [0, t_f]\}.$$

Finally note that the convex structure of an infinite-dimensional minimization problem allows us to write necessary and sufficient optimality conditions for a convex OCP (1.2a) in the form of the *Karush-Kuhn-Tucker* Theorem or in the form of an easy *variational inequality* (see [92, 120]).

4.4 Some Open Problems

The investigation of convex control systems involves a question of general interest. Let Ξ be a space of functions from $\mathbb{R}$ into $\mathbb{R}$. Let Π be a normed linear space and Θ be a topological space. Consider a mapping $\mathcal{T} : \Xi \times \Theta \rightarrow \Pi$ and assume that for every $\theta \in \Theta$ the given equation

$$\mathcal{T}(\xi, \theta) = 0_\Pi \tag{4.9}$$

has a unique solution $\xi^\theta(\cdot) \in \Xi$. It is a familiar consideration in applied mathematics to seek to solve (4.9) for ξ, while viewing θ as a parameter. Evidently, (4.9) can be interpreted as a generalization of a standard

control system. Related to the above theory of convex control systems
we can formulate the problem of finding constructive conditions for the
mapping $\mathcal{T}$ such that the functional

$$\mathcal{V} : \Theta \to \mathbb{R},\ \mathcal{V}(\theta) := \xi^\theta(t)$$

is convex for every $t \in \mathbb{R}$. Our main convexity results and the corresponding examples present a possible solution to the formulated gen-
eral problem for $\Theta = \mathcal{U}$ and the special case where the mapping $\mathcal{T}$ is
defined by ordinary differential equations.

The concept of convex optimal control problems my be of interest in
model predictive control when a sequence of open-loop optimal control
problems is solved to obtain feedback. Stability of the closed-loop cru-
cially depends on the fact that global optima for the involved open-loop
problems are found. The latter is clearly facilitated if the involved problems are convex. It seems also plausible that one can use the theoretical
results and algorithms presented in this paper for some optimal control
processes governed by partial differential equations.

Finally, note that the convex control systems introduced in Theorem 20
belong to the class of *monotone control systems* proposed by Angeli and
Sontag (see [8, 215]). The general interrelation between the convex and
monotone control systems are also of mathematical interest.

Chapter 5

Relaxed Problems

In this chapter we gather results on the approximation technique for relaxed (extended) optimal control problems proposed by the author. We examine control processes governed by ordinary differential equations in the presence of state, target and integral constraints.

5.1 Elements in Relaxation Theory

It is common knowledge that the initial optimal control problem does not always have a solution (see e.g., [98, 216, 94]). On the other hand, the corresponding relaxed problem has, under mild assumptions, an optimal solution [76, 105, 189]. This solution can be considered, in practice, for constructing an approximating solution for the initial problem. In the absence of the so-called *relaxation gap* (see e.g., [159, 160, 72]) the relaxed problem is of prime interest for the initial optimal control problem. In this case the minimal value of the objective functional in the initial problem coincides with the minimum of the objective functional in the relaxed problem. Therefore, in this situation a solution of the relaxed problem can be used as a basis for constructing a minimizing sequence for the initial problem [207, 94].

Extensions in problems of variational calculus, beginning with the idea of Hilbert, were realized many times for various purposes. Let us consider the regular variational problem

$$\int_{\Upsilon} \mathcal{L}(s, \omega(s), \dot{\omega}(s)) ds \rightarrow \min$$
$$\omega = \varsigma \text{ on } \partial \Upsilon$$

In the 20th problem of his Paris lecture, Hilbert asked the following question: "Is it true that the presented variational problem has a solution $\omega(\cdot)$ in case we generalize the notion of solution in an appropriate sense?" Today we know a positive answer. Under some natural assumption (see e.g., [221]) for a bounded region Υ in $\mathbb{R}^n$ and for the smooth functions $\mathcal{L}(\cdot, \cdot, \cdot)$, $\varsigma(\cdot)$ there exists a solution $\omega(\cdot)$ from the Sobolev space $\mathbb{W}_l^1(\Upsilon)$, $1 < l < \infty$. The first constructive investigation on this subject was a work of N.N. Bogoljubov [53]. The concept of relaxed controls was introduced by L.C. Young in 1937 under the name of generalized curves and surfaces [219]. It has been used extensively in the literature for the study of diverse optimal control problems [216, 76, 106, 120, 94, 189]. As Clarke points out in [80], a relaxed problem is, in general, the only one for which existence theorems can be proved and, for this reason, there are many who deem it the only reasonable problem to consider in practice. However, though relaxation of a problem is important in order to prove existence theorems, one is also interested in proving convergence of numerical approximation algorithms.

In [19, 24, 27] we consider an OCP of the type (1.2a) with a given terminal functional $\mathcal{J}(u(\cdot)) := \phi(x^u(1))$ and with an additional integral constraint

$$\int_0^{t_f} s(t, x(t), u(t))dt \leq 0,$$

where $s : [0, t_f] \times \mathbb{R}^n \times \mathbb{R}^m \to \mathbb{R}$ is a continuous function. This way we incorporate in our optimization framework (1.4a) the additional inequality $\tilde{s}(u(\cdot)) \leq 0$, where

$$\tilde{s}(u(\cdot)) := \int_0^{t_f} s(t, x^u(t), u(t))dt.$$

Let $\mathbb{L}_1^1([0, t_f])$ be the standard Lebesgue space of all integrable functions. We use the following notation

$$\aleph(n+1) := \{(\alpha^1(\cdot), ..., \alpha^{n+1}(\cdot))^T \ : \ \alpha^j(\cdot) \in \mathbb{L}_1^1([0, t_f]), \ \alpha^j(t) \geq 0,$$

$$\sum_{j=1}^{n+1} \alpha^j(t) = 1 \ \forall t \in [0, t_f]\}, \ \alpha(\cdot) := (\alpha^1(\cdot), ..., \alpha^{n+1}(\cdot))^T.$$

In parallel with the given control system (1.3) we consider the relaxed control system in the form of *Gamkrelidze system* (*Gamkrelidze chattering*) [106, 94]

$$\dot{y}(t) = \sum_{j=1}^{n+1} \alpha^j(t) f(t, y(t), u^j(t)) \quad \text{a.e. on } [0, t_f],$$

$$y(0) = x_0,$$

$$\tag{5.1}$$

where $\alpha(\cdot) \in \aleph(n + 1)$ and $u^j(\cdot) \in \mathcal{U}$, $j = 1, ..., n + 1$. Under the assumptions of Section 1.2 there exists an absolutely continuous solution $y^v(\cdot)$ of the initial value problem (5.1) associated with an admissible *generalized control*

$$v(\cdot) \in \aleph(n + 1) \times \mathcal{U}^{n+1},$$

where $v(t) := (\alpha^1(t), ..., \alpha^{n+1}(t), u^1(t), ..., u^{n+1}(t))^T$. Recall that *Radon probability measure* ς on the Borel sets of U is a regular positive measure ς such that $\varsigma(U) = 1$. Let $\mathcal{M}^1_+(U)$ be the space of all probability measures on the Borel sets of U. A *relaxed control*, $\mu(\cdot)$ is a measurable function $\mu : [0, t_f] \to \mathcal{M}^1_+(U)$, where "measurability" is defined in [216]. Following [94] we denote by $R_c([0, t_f], U)$ the set of relaxed controls. These controls give rise to the relaxed dynamics

$$\dot{\eta}(t) = \int_U f(t, \eta(t), u)\mu(t)(du) \text{ a.e on } [0, t_f], \tag{5.2}$$
$$\eta(0) = x_0.$$

The considered control problem can also be rewritten as a control problem for the differential inclusion

$$\dot{x}(t) \in f(t, x(t), U) \text{ a.e. on } [0, t_f],$$
$$x(0) = x_0,$$

where

$$f(t, x(t), U) := \{f(t, x(t), u) \ : \ u \in U\}.$$

Under the above-mentioned standard assumptions a function $x(\cdot)$ is a solution of the initial control system (1.3) with $u(\cdot) \in \mathcal{U}$ if and only if it is a solution of the initial differential inclusion given above. For details, see [94]. Related to the initial differential inclusion we consider the *relaxed differential inclusion*

$$\dot{x}(t) \in \text{conv} f(t, x(t), U) \text{ a.e. on } [0, t_f],$$
$$x(0) = x_0. \tag{5.3}$$

Let us now present the fundamental Equivalence Theorem [76, 216, 94].

Proposition 2 *Let the initial control system* (1.3) *satisfy the assumptions presented in* Section 1.2. *A function* $y(\cdot)$ *is a solution of the* Gamkrelidze *system* (5.1) *with* $v(\cdot) \in \aleph(n+1) \times \mathcal{U}^{n+1}$ *if and only if it is a solution* *of the relaxed differential inclusion (5.3). Moreover, a function $x(\cdot)$ is a solution of the relaxed differential inclusion (5.3) if and only if it is a solution of the relaxed problem (5.2) with a relaxed control $\mu(\cdot)$.*

Let

$$v(\cdot) = (\alpha^1(\cdot), ..., \alpha^{n+1}(\cdot), u^1(\cdot), ..., u^{n+1}(\cdot))^T,$$
$$\bar{v}(\cdot) = (\bar{\alpha}^1(\cdot), ..., \bar{\alpha}^{n+1}(\cdot), \bar{u}^1(\cdot), ..., \bar{u}^{n+1}(\cdot))^T.$$

As a consequence of the basic existence and uniquess theorems for differential equations (see e.g., [44, 178]) one can obtain the following useful estimates [178].

Proposition 3 *Let the initial control system (1.3) satisfy assumptions presented in* Section 1.2. *There exist finite constants c_1, c_2, c_3 and c_4 such that*

$$\|x^u(\cdot)\|_{\mathbb{L}_n^\infty([0,t_f])} \leq c_1, \; \|x^u(\cdot) - x^{\tilde{u}}(\cdot)\|_{\mathbb{L}_n^\infty([0,t_f])} \leq$$
$$\leq c_2 \|u(\cdot) - \tilde{u}(\cdot)\|_{\mathbb{L}_m^2([0,t_f])},$$
$$\|y^v(\cdot)\|_{\mathbb{L}_n^\infty([0,t_f])} \leq c_3, \; \|y^v(\cdot) - y^{\bar{v}}(\cdot)\|_{\mathbb{L}_n^\infty([0,t_f])} \leq$$
$$\leq c_4 \sum_{j=1}^{n+1} \|u^j(\cdot) - \bar{u}^j(\cdot)\|_{\mathbb{L}_m^2([0,t_f])},$$

for all $u(\cdot), \tilde{u}(\cdot) \in \mathcal{U}$ and for all $v(\cdot), \bar{v}(\cdot) \in \aleph(n+1) \times \mathcal{U}^{n+1}$, where $x^u(\cdot), x^{\tilde{u}}(\cdot)$ are the solutions of (1.3) and $y^v(\cdot), y^{\bar{v}}(\cdot)$ are the solutions of (5.1) associated with the controls $u(\cdot), \tilde{u}(\cdot)$ and $v(\cdot), \bar{v}(\cdot)$, respectively.

Using the solution $y^v(\cdot)$ of (5.1), we can define the relaxation of the corresponding minimization problem (1.4a)

minimize $\bar{\mathcal{J}}(v(\cdot))$

subject to $v(\cdot) \in \aleph(n+1) \times \mathcal{U}^{n+1}$ (5.4)

$\bar{h}(v(\cdot)) \leq 0, \; \bar{q}(v(\cdot))(t) \leq 0 \; \forall t \in [0, t_f], \; \bar{s}(v(\cdot)) \leq 0,$

where

$$\bar{\mathcal{J}}(v(\cdot)) := \phi(y^v(t_f)),$$
$$\bar{h}(v(\cdot)) := h(y^v(t_f)),$$
$$\bar{q}(v(\cdot))(t) := q(t, y^v(t)) \; \forall t \in [0, t_f],$$
$$\bar{s}(v(\cdot)) := \int_0^{t_f} \sum_{j=1}^{n+1} \alpha^j(t) s(t, y^v(t), u^j(t)) dt.$$

Note that under some mild assumptions the relaxed control problem
(5.4) has an optimal solution (see [105]). We denote an optimal solution
by $v^{opt}(\cdot)$.

Let $\mathbb{L}^1([0, t_f], \mathbb{C}(U))$ denote the space of absolutely integrable functions
from $[0, t_f]$ to $\mathbb{C}(U)$ (the space of all continuous function on U). The
topology imposed on $R_c([0, t_f], U)$ is the weakest topology such that the
mapping

$$\mu \to \int_0^{t_f} \int_U \psi(t, u)\mu(t)(du)dt$$

is continuous for all $\psi(\cdot, \cdot) \in \mathbb{L}^1([0, t_f], \mathbb{C}(U))$. Finally we recall ([216],
p.287) that $R_c([0, t_f], U)$ is a compact and convex space of the dual space
to $\mathbb{L}^1([0, t_f], \mathbb{C}(U))^*$. The topology of the "weak" norm of

$$\mathbb{L}^1([0, t_f], \mathbb{C}(U))^*$$

restricted to $R_c([0, t_f], U)$ coincides with the weak star topology [216].

5.2   β-Relaxations

A variety of approximation schemes have been recognized as a powerful tool for theoretical studying and practical solving the infinite dimensional optimization problems. On the other hand, theoretical approaches to the relaxed optimal control problem with constraints are not
sufficiently advanced to the numerically tractable schemes. Some possible approximations of the differential inclusion were examined from
the theoretical standpoint in [85, 157, 158, 159, 214]. As an example
we refer to the *proximal-like* method (in the state space) proposed by
Mordukhovich in [159, 160] for the Bolza problem. An implementation of this method is directly related to a sequence of sophisticated
optimization problems. Besides, it is difficult to generalize this method
to a relaxed optimal control problem with additional constraints.

For the constructive approximation of the relaxed control system we use
the generalized system of Gamkrelidze (5.1). Our aim is to investigate
(5.1) and the corresponding problem (5.4) from the viewpoint of numerically tractable approximations. For this purpose we introduce in
[19, 24, 27] so-called β-systems. At first, we approximate the control
set U by a finite set U_M of points

$$v^k \in U, \ k = 1, ..., M,$$

where $M = (n + 1)\tilde{M}$, $\tilde{M} \in \mathbb{N}$. We assume that for a given number $\epsilon > 0$ (accuracy) there exists a natural number $\tilde{M}_\epsilon$ such that for every $v \in U$ one can find a point $v^k \in U_M$ with $\|v - v^k\|_{\mathbb{R}^m} < \epsilon$, where $M = (n+1)\tilde{M}$, $\tilde{M} > \tilde{M}_\epsilon$. We now consider a sequence of the above accuracies $\{\epsilon_M\}$ such that $\lim_{M \to \infty} \epsilon_M = 0$. Let us introduce the following auxiliary system

$$\dot{z}(t) = \sum_{k=1}^{M} \beta^k(t) f(t, z(t), v^k) \text{ a.e. on } [0, t_f], \ z(0) = x_0, \qquad (5.5)$$

where $\beta^k(\cdot) \in \mathbb{L}_1^1([0, t_f])$, $\beta^k(t) \geq 0$ and $\sum_{k=1}^{M} \beta^k(t) = 1 \ \forall t \in [0, t_f]$. Define the set of admissible β-controls

$$\aleph(M) := \{(\beta^1(\cdot), ..., \beta^M(\cdot))^T \ : \ \beta^k(\cdot) \in \mathbb{L}_1^1([0, t_f]), \ \beta^k(t) \geq 0,$$

$$\sum_{k=1}^{M} \beta^k(t) = 1 \ \forall t \in [0, t_f]\}, \ \beta_M(\cdot) := (\beta^1(\cdot), ..., \beta^M(\cdot))^T.$$

We call the introduced control system (5.5) β-*system* and a function $\beta_M(\cdot) \in \aleph(M)$ a β-*control*. Note that for a fixed U_M the given β-system (5.5) has an absolutely continuous solution $z_M^\beta(\cdot)$ for every admissible β-control $\beta_M(\cdot) \in \aleph(M)$.

Let $\mathbb{W}_n^{1,1}([0, T])$ be the standard Sobolev space of absolutely continuous functions $\varphi : [0, t_f] \to \mathbb{R}^n$, $\dot{\varphi}(\cdot) \in \mathbb{L}_n^1([0, t_f])$. The function space $\mathbb{W}_n^{1,1}([0, t_f])$ equipped with the norm $\| \cdot \|_{\mathbb{W}_n^{1,1}([0,t_f])}$ defined by

$$\|\varphi(\cdot)\|_{\mathbb{W}_n^{1,1}([0,t_f])} := \|\varphi(\cdot)\|_{\mathbb{L}_n^1([0,t_f])} + \|\dot{\varphi}(\cdot)\|_{\mathbb{L}_n^1([0,t_f])} \ \forall \varphi(\cdot) \in \mathbb{W}_n^{1,1}([0, t_f])$$

is a Banach space. By $\mathbb{C}_n([0, t_f])$ we denote the Banach space of all continuous functions $\varrho : [0, t_f] \to \mathbb{R}^n$ equipped with the usual max-norm

$$\|\varrho(\cdot)\|_{\mathbb{C}_n([0,t_f])} := \max_{t \in [0,t_f]} \|\varrho(t)\|_{\mathbb{R}^n}.$$

In [19, 24, 27] we first establish the following basic properties of the introduced β-systems.

Theorem 25 *Let the right-hand side of system* (1.3) *satisfy the assumptions of* Proposition 1. *For every*

$$(v(\cdot), y^v(\cdot)), \ v(\cdot) \in \aleph(n + 1) \times \mathcal{U}^{n+1}$$

there exists a sequence of β-controls $\{\beta_M(\cdot)\} \subset \aleph(M)$ and the corresponding sequence $\{z_M^\beta(\cdot)\}$ of solutions of the appropriate β-systems such that $z_M^\beta(\cdot)$ approximate the solution $y^\nu(\cdot)$ of (5.1) in the following sense

$$\lim_{M\to\infty} \|z_M^\beta(\cdot) - y^\nu(\cdot)\|_{\mathbb{C}_n([0,t_f])} = 0$$

and

$$\lim_{M\to\infty} \|z_M^\beta(\cdot) - y^\nu(\cdot)\|_{\mathbb{W}_n^{1,1}([0,t_f])} = 0.$$

Theorem 26 *Let the right-hand side of system* (1.3) *satisfy the assumptions of* Proposition 1. *Assume that*

$$\beta_M(\cdot) \in \aleph(M)$$

and $z_M^\beta(\cdot)$ is the corresponding solution of (5.5). *Then*

$$\dot{z}_M^\beta(t) \in \operatorname{conv}\{f(t, z_M^\beta(t), U)\} \text{ a.e. on } [0, t_f].$$

Theorem 25 and Theorem 26 show that $z_M^\beta(\cdot)$ is an approximation to the solution of the relaxed differential inclusion (5.3).

Theorem 27 *Let the right-hand side of system* (1.3) *satisfy the assumptions of* Proposition 1. *Let*

$$\beta_M(\cdot) \in \aleph(M)$$

and $z_M^\beta(\cdot)$ be the corresponding solution of (5.5). *Then there exists a piecewise constant control function $\tilde{u}_M(\cdot) \in \mathcal{U}$ such that the solution $x^{\tilde{u}_M}(\cdot)$ of the initial system* (1.3) *exists in $[0, t_f]$ and*

$$\lim_{M\to\infty} \|z_M^\beta(t) - x^{\tilde{u}_M}(t)\|_{\mathbb{R}^n} = 0$$

uniformly in $t \in [0, t_f]$.

The presented theorems make it possible to approximate the initial problem (5.4). Let us consider the functions $\bar{\mathcal{J}}_M, \bar{h}_M, \bar{s}_M : \aleph(M) \to \mathbb{R}$ and

$$\bar{q}_M : \aleph(M) \to \mathbb{C}([0, t_f])$$

defined by

$$\bar{\mathcal{J}}_M(\beta_M(\cdot)) := \phi(z_M^\beta(t_f)), \ \bar{h}_M(\beta_M(\cdot)) := h(z_M^\beta(t_f)),$$
$$\bar{q}_M(\beta_M(\cdot))(t) := q(t, z_M^\beta(t)) \ \forall t \in [0, t_f],$$
$$\bar{s}_M(\beta_M(\cdot)) := \int_0^{t_f} \sum_{k=1}^M \beta^k(t) s(t, z_M^\beta(t), v^k) dt.$$

We now can present our next result from [19, 24, 27].

Theorem 28 *For every $\delta > 0$ there exists a number $M_\delta \in \mathbb{N}$ such that for all natural numbers $M > M_\delta$ the following β-relaxed optimal control problem*

$$\begin{aligned} &\text{minimize } \bar{\mathcal{J}}_M(\beta_M(\cdot)) \text{ subject to } \beta_M(\cdot) \in \aleph(M), \\ &\bar{h}_M(\beta_M(\cdot)) \le \delta, \\ &\bar{q}_M(\beta_M(\cdot))(t) \le \delta \ \forall t \in [0, t_f], \\ &\bar{s}_M(\beta_M(\cdot)) \le \delta, \end{aligned} \tag{5.6}$$

has an optimal solution $\beta_M^{opt}(\cdot) \in \aleph(M)$ on condition that (5.6) has an admissible solution. Moreover,

$$|\bar{\mathcal{J}}_M(\beta_M^{opt}(\cdot)) - \bar{\mathcal{J}}(v^{opt}(\cdot))| \le \delta,$$

where $v^{opt}(\cdot) \in \aleph(n+1) \times \mathcal{U}^{n+1}$ is an optimal solution of the relaxed optimal control problem (5.4).

The β-system (5.5) and the corresponding β-relaxed optimization problem (5.6) provided a basis for the numerical algorithms for relaxed optimal control problems (5.4). Here we replace the given inequalities constraints from (5.4) by the corresponding weakened inequalities.

5.3 Numerical Aspects

Suppose that we have a sequence of optimal β-controls

$$\{\beta_M^{opt}(\cdot)\}, \; M \in \mathbb{R}$$

for problems (5.6). Using this sequence, we can define the associated minimizing sequence of ordinary controls from $\mathcal{U}$ in a modified Tichomirov-way [207]. The concrete mathematical constructions of a sequence of this sort is presented in [24, 27]. We define a *Tichomirov-like* sequence $\{u_{\Omega,M}(\cdot)\}$, $\Omega \in \mathbb{N}$ of piecewise constant control functions from $\mathcal{U}$. By our definition, $u_{\Omega,M}(t) \in U_M$ for all $t \in [0, t_f]$. In the above-mentioned works [24, 27] we deduce the following two approximating properties of the Tichomirov-like sequence $\{u_{\Omega,M}(\cdot)\}$ with respect to the β-system (5.5) and with respect to the optimization problem (5.6).

Theorem 29 *Let* $\{u_{\Omega,M}(\cdot)\}$ *be the* Tichomirov-like *sequence generated by* $\{\beta_M^{opt}(\cdot)\}$. *Then*

$$\lim_{\Omega \to \infty} \|z_M^{\beta,opt}(\cdot) - x^{u_{\Omega,M}}(\cdot)\|_{\mathbb{C}_n([0,t_f])} = 0,$$

where $z_M^{\beta,opt}(\cdot)$ *is the solution of* (5.5) *associated with* $\beta_M^{opt}(\cdot)$ *and trajectory* $x^{u_{\Omega,M}}(\cdot)$ *is the solution of* (1.3) *corresponding to the control* $u_{\Omega,M}(\cdot)$.

Theorem 30 *Let* $\{u_{\Omega,M}(\cdot)\}$ *be the* Tichomirov-like *sequence generated by* $\{\beta_M^{opt}(\cdot)\}$. *Then for a fixed* $M \in \mathbb{N}$ *and for every* $\Delta > 0$ *there exists a number* $\Omega_\Delta \in \mathbb{N}$ *such that*

$$|\mathcal{J}(u_{\Omega,M}(\cdot)) - \bar{\mathcal{J}}_M(\beta_M^{opt}(\cdot))| \le \Delta,$$

and

$$\tilde{h}(u_{\Omega,M}(\cdot)) \le \Delta + \delta,$$
$$\tilde{q}(u_{\Omega,M}(\cdot))(t) \le \Delta + \delta \; \forall t \in [0, t_f],$$
$$\tilde{s}(u_{\Omega,M}(\cdot)) \le \Delta + \delta.$$

for all natural numbers $\Omega > \Omega_\Delta$.

It is necessary to stress that the elements of the Tichomirov-like sequence $\{u_{\Omega,M}(\cdot)\}$ usually are not admissible solutions of the minimization problem (1.4a). If we want to apply the above results to the initial problem (1.4a), we need the following concept [159, 160].

Definition 8 *The initial* OCP (1.4a) *is called stable with respect to relaxation* (5.4), *if*

$$\inf_{Problem(1.4a)} \mathcal{J}(u(\cdot)) = \min_{Problem(5.4)} \bar{\mathcal{J}}(v(\cdot)).$$

An analogous phenomenon of *relaxation gaps* is considered in [72]. For example, the relaxation stability holds for one-dimensional optimal control problems (1.2) without state and integral constraints [158, 159]. In general, the relaxation stability is related to the property of *calmness* [78, 79]. Note that according to Clarke [79], the calmness hypothesis implies that corresponding necessary optimality conditions can be taken as *normal*. A general result that normality implies relaxation stability for a class of optimal control problems has been obtained by Warga [216, 217]. For special classes of problems (1.2) without state and integral constraints the relaxation stability holds with no calmness or normality assumptions. If we assume that the initial OCP of the type (1.2a) with a given terminal functional $\mathcal{J}(u(\cdot)) := \phi(x^u(1))$ and with an additional integral constraint is stable with respect to relaxation (5.4), we can formulate our main result from [24, 27].

Theorem 31 *Let the considered* OCP *be stable with respect to relaxation* (5.4). *The* Tichomirov-like *sequence* $\{u_{\Omega,M}(\cdot)\}$ *generated by* $\{\beta_M^{opt}(\cdot)\}$ *is a minimizing sequence for the corresponding initial optimization problem* (1.4). *That means that for every* $\epsilon > 0$ *there exist* $M_\epsilon \in \mathbb{N}$ *such that for all* $M > M_\epsilon$ *one can find a number* $\Omega_\epsilon(M) \in \mathbb{N}$ *with*

$$|\mathcal{J}(u_{\Omega,M}(\cdot)) - \inf_{Problem(1.4a)} \mathcal{J}(u(\cdot))| \leq \epsilon$$

$$\tilde{h}(u_{\Omega,M}(\cdot)) \leq \epsilon,$$
$$\tilde{q}(u_{\Omega,M}(\cdot))(t) \leq \epsilon \ \forall t \in [0, t_f],$$
$$\tilde{s}(u_{\Omega,M}(\cdot)) \leq \epsilon$$

for all $\Omega > \Omega_\epsilon(M)$, $M > M_\epsilon$.

Theorem 31 shows that in the case of stability with respect to relaxation (5.4) the solutions of the β-relaxed optimal control problems (5.6) can be used for constructing a minimizing sequence for the considered initial optimal control problem (1.4). To demonstrate the feasibility and effectiveness of solving relaxed control problems via β-relaxation, we examine two illustrative examples. Let us introduce a grid

$$G_N := \{t_1, t_2, ..., t_{N+1}\},$$
$$t_1 = 0, \ t_{N+1} = t_f, \ t_{r+1} = t_r + t_f/N, \ r = 1, ..., N.$$

For a number $M \in \mathbb{N}$ we now examine a finite-dimensional variant of problem (5.6)

$$
\begin{aligned}
&\text{minimize } \bar{\mathcal{J}}_M(\beta_M^N(\cdot)) \\
&\text{subject to } \beta_M^N(\cdot) \in \aleph(M) \bigcap \mathbb{L}_1^{2,N}(G_N), \\
&\bar{h}_M(\beta_M^N(\cdot)) \leq \delta, \\
&\bar{q}_M(\beta_M^N(\cdot))(t) \leq \delta \ \forall t \in [0, t_f], \\
&\bar{s}_M(\beta_M^N(\cdot)) \leq \delta,
\end{aligned}
\tag{5.7}
$$

where

$$\beta_M^N(\cdot) = (\beta^{1,N}(\cdot), ..., \beta^{M,N}(\cdot))$$

and $\beta^{k,N}(\cdot) \in \mathbb{L}_1^{2,N}(G_N)$. By $\mathbb{L}_1^{2,N}(G_N)$ we denote here the finite-dimensional Hilbert space of piecewise constant functions. Moreover, we consider a discretization of β-system (5.5). We use the Euler method for this purpose

$$z_M^N(t_{r+1}) = z_M^N(t_r) + \frac{t_f}{N} \sum_{k=1}^{M} \beta^{k,N}(t_r) f(t_r, z_M^N(t_r), v^k), \ r = 1, ..., N,$$
$$z_M^N(t_1) = x_0.$$
$$\tag{5.8}$$

It must be admitted that the above approximations of β-controls and the first order Euler discretizations (5.8) of β-system (5.5) are particularly advantageous for relatively easy control systems. Some alternative approximation procedures are described in [214, 178]. The selection of numerical methods for solving the finite-dimensional optimization problem (5.7) is arbitrary (in some sense). The difficult question, how this discretization procedure influences the overall precision and the convergence properties of the above approximations is of major importance. There are a number of publications devoted to numerical aspects of discrete approximations for optimal control problems. We refer to [159, 160, 85, 149, 88, 89, 115, 116] for theoretical details. For example, the convergence of Euler-type approximations for Mayer-type optimal control problems is established in [82]. An abstract framework for proving convergence of discrete approximations is developed in [83]. For convergence analysis of some algorithms involving discrete approximations and for the corresponding survey see also [90, 131, 174, 175]. Let us now give a brief summary of the algorithm. For purposes of calculation, we use the gradient-type method (see Section 3.3 and Section 3.4). The reduced gradient for the discrete unconstrained problem (5.7) can be computed as follows

$$\nabla_\beta \bar{\mathcal{J}}_M(t_r) := -\frac{\partial}{\partial \beta} \mathcal{H}(z_M^N(t_r), \beta_M^N(t_r), p(t_{r+1})), \ r = 1, ..., N,$$

where $z_M^N(\cdot)$ is the solution of (5.8) and $\mathcal{H}$ is the Hamiltonian for (5.7)

$$\mathcal{H}(z_M^N(t_r), \beta_M^N(t_r), p(t_{r+1})) := \langle p(t_{r+1}), z_M^N(t_r)+$$
$$\frac{t_f}{N} \sum_{k=1}^{M} \beta^{k,N}(t_r) f(t_r, z_M^N(t_r), v^k) \rangle_{\mathbb{R}^n}, \ r = 1, ..., N$$

The adjoint variable $p(\cdot)$ is given by

$$p(t_r) = \frac{\partial}{\partial z} \mathcal{H}(z_M^N(t_r), \beta_M^N(t_r), p(t_{r+1})), \ r = 1, ...N,$$

$$p(t_{N+1}) = -\frac{\partial}{\partial z} \phi(z_M^N(t_{N+1})).$$

Let us now present the Projected Gradient Algorithm for the discrete problem (5.7) (see e.g., [100, 172, 173])

$$\beta_M^{N,(l+1)}(t_r) = P_{\mathcal{V}_\beta}(\beta_M^{N,(l)}(t_r) - \gamma_l \nabla_\beta \bar{\mathcal{J}}_M(t_r)),$$
$$l = 0, 1, ...$$

where γ_l is a step-size and $P_{\mathcal{V}_\beta}$ is a projection on the set

$$\mathcal{V}_\beta := \{\beta \in \aleph(M) \bigcap \mathbb{L}_1^{2,N}(G_N) \ : \ \bar{h}_M(\beta) - \delta \le 0,$$
$$\bar{q}_M(\beta)(t) - \delta \le 0 \ \forall t \in [0, t_f], \ \bar{s}_M(\beta) - \delta \le 0\}.$$

Example 9 *Consider the following convex* OCP *(see [24, 27])*

$$\text{minimize } x^2(1)$$
$$\text{subject to } \dot{x}(t) = u(t), \ x(0) = 0,$$
$$|u(t)| \le 1, \ u(\cdot) \in \mathbb{L}^1([0, 1]), \tag{5.9}$$
$$1/4 - \int_0^1 x(t)dt \ge 0,$$

and apply the proposed techniques of β-relaxation and the proximal point algorithm. Evidently, this example needs no relaxation. However, problem (5.9) can be considered as a possible test for the presented approximation schemes.

The set of admissible solutions of problem (5.9) is nonempty. For instance, $u(t) = 1$ is an admissible solution. The following bang-bang control

$$u^{opt}(t) = \begin{cases} 1 & \text{if } t \in [0, 0.5), \\ -1 & \text{if } t \in [0.5, 1]. \end{cases}$$

is a unique optimal solution of (5.9). Let $M = 51$ and $N = 50$. Clearly,

$$U_M := \{-1.00, -0.96, ..., 0.96, 1.00\}$$

and

$$G_N := \{0.00, 0.02, ..., 0.98, 1.00\}.$$

The β-system and the corresponding Euler-discretizations *(5.8) for (5.9) are*

$$\dot{z}(t) = \sum_{k=1}^{51} \beta^k(t)v^k,$$

$$z_M^N(t_{r+1}) = z_M^N(t_r) + 0.02 \sum_{k=1}^{51} \beta^{k,N}(t_r)v^k,$$

where

$$v^k \in U_M, \ r = 1, ..., N.$$

The Tichomirov-like *ordinary control $u_{\Omega,M}(\cdot)$ can be computed for an arbitrary natural number $\Omega \ge 2$. This control is of the bang-bang type (Fig. 5.1).*

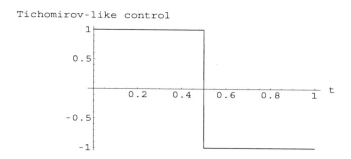

Figure 5.1: Optimal Tichomirov-like control, $u_{\Omega,M}(t)$

For the purpose of simple numerical computations we assume that the switching point $t^{sw} = 0.5$ (the critical jump of the optimal control) is included into the grid G_N. Here we present only one discrete solution. The optimal control problem (5.9) was also solved with the efficient direct collocation technique DIRCOL (see [201]). Note that both of these computational results for optimal control coincide here very closely. We apply the classical proximal point algorithm to problem (5.9) with the following positive parameters

$$\chi_\lambda = 1 + 1/2^\lambda, \ \lambda \in \mathbb{N}.$$

Evidently, a particular computational result essentially depends on the selection of the sequence $\{\chi_\lambda\}$. Note that a large parameter χ_λ can involve some loss of precision.

Example 10 *Our next problem is the following* Goddard-type *problem of rocket vertical ascent (see [109] and [27])*

minimize $- x_2(100)$

subject to $\dot{x}_1(t) = -u(t), \ \dot{x}_2(t) = x_3(t),$

$$\dot{x}_3(t) = -0.01 + \frac{2u(t) - 0.05x_3^2(t)\exp(-0.1x_2(t))}{x_1(t)},$$

$x_1(0) = 1,$ (5.10)

$x_2(0) = x_3(0) = 0,$

$x_1(100) = 0.2,$

$0 \leq u(t) \leq 0.04, \ t \in [0, 100],$

$u(\cdot) \in \mathbb{L}^1([0, 100]),$

where $x_1(t)$, $x_2(t)$ and $x_3(t)$ are the mass, the altitude and the vertical velocity of the rocket, respectively. The original Goddard *problem has been treated theoretically and numerically by many authors (see e.g., [155, 153]). This famous problem differs slightly from problem (5.10). Originally, the final time t_f is free and, in addition, $x_3(t_f) = 0$ is prescribed.*

Our aim is to maximize the attained altitude $x_2(100)$ of the rocket. Moreover, the terminal mass should be equal to 0.2. For M, N, U_M and G_N from Example 9 *we apply the proximal-based computational method to the β-relaxation of the optimal control problem (5.10). The* Tichomirov-like *control is shown in* Fig. 5.2.

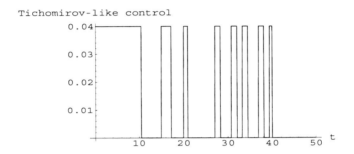

Figure 5.2: Optimal Tichomirov-like control, $u_{\Omega,M}(t)$

For comparison the Goddard-type *problem (5.10) was also solved using the direct collocation method and the DIRCOL package [201]. The calculated mass, altitude and vertical velocity of the rocket are shown in Fig. 5.3, Fig. 5.4 and Fig. 5.5.*

In all figures the dotted line denotes the computational results obtained by DIRCOL. The corresponding results for the β-relaxation are presented by continuous lines. The constraints are satisfied with tolerance 10^{-5}. Using the technique based on the β-relaxation, we obtain the following computed optimal objective value: -131.901. The optimal objective value calculated by the DIRCOL code is equal to -132.1961. The presented computational results accord well with the second convergence result of Theorem 25. *This is especially true in regard to the large parameters M and N of the corresponding grids.*

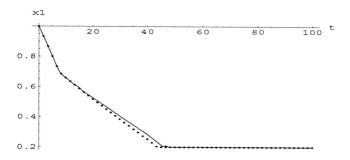

Figure 5.3: Optimal mass of the rocket

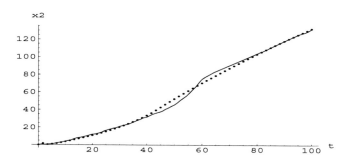

Figure 5.4: Optimal altitude of the rocket

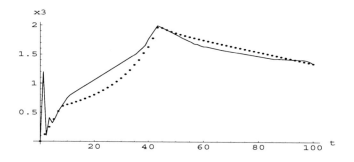

Figure 5.5: Optimal vertical velocity of the rocket

5.4 On Approximately Convex OCPs

The considered β-system (5.5) has a linear structure with respect to the new control functions, namely, to the β-controls $\beta_M(\cdot) \in \aleph(M)$. Moreover, the set of admissible controls $\aleph(M)$ is convex. Let us examine the relaxed OCP of the type (1.2a) with a terminal functional $\mathcal{J}$ and with the semilinear state equation (1.3). The relaxed dynamics is described by

$$\dot{x}(t) = A(t)x(t) + \int_U B(t, u)\mu(t)(du) \text{ a.e. on } [0, t_f],$$

$$x(0) = x_0.$$

In this case the corresponding optimization problem is a convex minimization problem in a very sophisticated Banach space of all relaxed controls, namely, in the space $(\mathbb{L}^1([0, t_f], \mathbb{C}(U)))^*$ (see [26], Theorem 2, pp.9-10). On the other hand, the associated β-system

$$\dot{z}(t) = A(t)z(t) + \sum_{k=1}^{M} \beta^k(t)B(t, v^k) \text{ a.e. on } [0, t_f],$$

$$z(0) = x_0$$

is linear with respect to the vector of new controls $\beta_M(\cdot) \in \aleph(M)$. Using the main results of this chapter, we can prove the following result.

Theorem 32 *Let the considered OCP be stable with respect to relaxation (5.4). If the functionals $\phi(\cdot)$, $h(\cdot)$ are convex and the functional $q(t, \cdot)$ is convex for all $t \in [0, t_f]$, then the corresponding β-relaxed problem (5.6) is a convex optimal control problem and the given OCP (1.2a) is an approximately convex problem.*

In a similar manner, we obtain our next example of an approximately convex OCP ([26], pp.13-14).

Theorem 33 *Consider the control system (1.3) with*

$$f(t, x, u) = f^1(t, x) + f^2(t, u),$$

where f^1, f^2 are a continuous function and for all $t \in [0, t_f]$

$$\|f^1(t, x_1) - f^1(t, x_2)\| \leq L\|x_1 - x_2\|, \quad \forall x_1, x_2 \in \mathbb{R}^n.$$

Let $f_k^1(t, \cdot)$, $f_k^2(t, \cdot)$ $k = 1, ..., n$ be convex and monotonically nondecreasing functionals for every $t \in [0, t_f]$. Let the considered OCP be stable with respect to relaxation (5.4). Assume that the functionals $\phi(\cdot)$, $h(\cdot)$ are convex, monotonically nondecreasing and the function $q(t, \cdot)$ is convex, monotonically nondecreasing for every $t \in [0, t_f]$. Then the corresponding β-relaxed problem is a convex optimal control problem and the given OCP (1.2a) is an approximately convex problem.

Finally note that control systems from Theorem 32 and Theorem 33 generalize the convex systems introduced in the previous chapter.

The proposed technique of β-relaxations can also be considered in the context of the numerical approximation of *Young measures* (see [189, 190]). In distinction to the method of control parametrization [205], our method makes the use of a variety of optimization algorithms for solving the auxiliary β-relaxed problem possible. As noted above (see Example 9), the used proximal-based method has a drawback, namely, a particular computational result essentially depends on the selection of the sequence of parameters χ (see also the computational results of [19]). Generally speaking, the presented numerical approximation of the control set U is not without disadvantages. The prime interest here is with a possible effective approximations of the *orientor field* $f(t, x(t), U)$.

Chapter 6

Special Systems and Problems

This chapter contains some results for a class of hybrid optimal control problems. Moreover, we also investigate a family of nonlinear optimal control problems, which can be called "optimal control problems in mechanics". We propose computational scheme for the above classes of problems and consider some illustrative examples.

6.1   On Hybrid Optimal Control Problems

Hybrid control systems are mathematical modes of heterogeneous systems consisting of a continuous part, a finite number of continuous controllers and a discrete supervisor. For a hybrid optimal control problem, the main tool toward the construction of optimal trajectories is the Hybrid Maximum Principle [74, 170, 171, 196, 202, 32]. This result generalizes the classical Pontryagin Maximum Principle [44, 120, 176]. It is well-known that the standard proof of the Pontryagin Maximum Principle is based on the techniques of "needle variations" (see e.g., [120, 176]). The character of a general hybrid optimal control problem changes the possibility of using the standard needle variations [171]. Therefore, a variant of the Hybrid Maximum Principle for a hybrid optimal control problem can be proved only under some restrictive assumptions (see e.g., [170, 171, 196, 202]). In effect, these assumptions guarantee that the classical needle variations are still admissible variations. Furthermore, in the context of a practical implementation of the Hybrid Maximum Principle we need to construct a simultaneous solution of a large-dimensional boundary-value problem and of a family of sophisticated auxiliary minimization problems. This is a complicated problem, especially in the case of many-dimensional systems with a lot of switchings.

In this section, we consider a class of non-stationary hybrid control systems with autonomous (uncontrolled) location transitions. For general theory of hybrid systems and basic definitions we refer to, e.g.,

[161, 179]. Using an approach based on Lagrange-type techniques and on reduced gradients, we obtain a set of first-order necessary optimality conditions for the above class of nonlinear hybrid optimal control problems. The explicit computation of the corresponding reduced gradients provides also a basis for applications of some effective gradient-based optimization algorithms to the given hybrid optimal control problems. Let us first introduce the following variant of the standard definition of hybrid systems [170, 171, 196, 202, 31, 32].

Definition 9 *A hybrid system is a 7-tuple*

$$\{Q, M, U, F, \mathcal{U}, I, S\},$$

where

- *Q is a finite set of discrete states (called locations);*

- *$M = \{M_q\}_{q \in Q}$ is a family of smooth manifolds, indexed by Q;*

- *$U \subseteq \mathbb{R}^m$ is a set of admissible control input values (called control set);*

- *$F = \{f_q\}$, $q \in Q$ is a family of maps*

$$f_q : [0, 1] \times M_q \times U \to T M_q,$$

where $T M_q$ is the tangent bundle of M_q (see e.g., [80, 120]);

- *$\mathcal{U}$ is the set of all admissible control functions;*

- *$I = \{I_q\}$ is a family of subintervals of $[0, 1]$ such that the length of each I_q is less than 1;*

- *S is a subset of Ξ, where*

$$\Xi := \{(q, x, q', x') \; : \; q, q' \in Q, x \in M_q, x' \in M_{q'}\}$$

A hybrid system from Definition 9 is defined on the time-interval $[0, 1]$. Note that in contrast to the general definition of a hybrid system [170, 171, 196, 202], the control set U from Definition 9 is the same for all locations. Moreover, in the sense of this definition the set $\mathcal{U}$ is also independent of a location. Let us assume that U is a compact set and

$$\mathcal{U} := \{u(\cdot) \in \mathbb{L}^2_m([0,1]) \; : \; u(t) \in U \text{ a.e. on } [0,1]\},$$

where $\mathbb{L}^2_m([0,1])$ is the standard Lebesgue space of all square integrable functions $u : [0,1] \to \mathbb{R}^m$. We now introduce some additional hypothesis for the vector fields f_q, $q \in Q$:

- all functions $f_q(t, \cdot, \cdot)$ from F are differentiable,

- f_q, $\partial f_q/\partial x$, $\partial f_q/\partial u$ are continuous and there exist constants C_q such that

$$\|\frac{\partial}{\partial x} f_q(t, x, u)\| \leq C_q,$$
$$q \in Q, \; (t, x, u) \in [0,1] \times M_q \times U.$$

For $q, q' \in Q$ one can also define the *switching set*

$$S_{q,q'} := \{(x, x') \in M_q \times M_{q'} \; : \; (q, x, q'x') \in \mathcal{S}\}.$$

from location q to location q'. The intervals I_q, $q \in Q$ indicate the lengths of time intervals on which the system can stay in location q. We say that a location switching from q to q' occurs at a *switching time* $t^{switch} \in [0,1]$. We now consider a hybrid system with $r \in \mathbb{N}$ switching times $\{t_i\}$, $i = 1, ..., r$, where

$$0 = t_0 < t_1 < ... < t_r < t_{r+1} = 1.$$

Note that the sequence of switching times $\{t_i\}$ is not defined a priory. A hybrid control system remains in location $q_i \in Q$ for all

$$t \in [t_{i-1}, t_i[, \; i = 1, ..., r+1.$$

Let $t_i - t_{i-1} \in I_{q_i}$ for all $i = 1, ..., r+1$. A hybrid system (in the sense of Definition 9) that satisfies the above assumptions is denoted by $\mathcal{HS}$.

Definition 10 *Let $u(\cdot) \in \mathcal{U}$ be an admissible control for a hybrid control system $\mathcal{HS}$. Then a continuous trajectory of $\mathcal{HS}$ is an absolutely continuous function*

$$x : [0,1] \to \bigcup_{q \in Q} M_q$$

such that $x(0) = x_0 \in M_{q_1}$ and

- $\dot{x}(t) = f_{q_i}(t, x(t), u(t))$ *for almost all* $t \in [t_{i-1}, t_i]$ *and all*

$$i = 1, ..., r + 1;$$

- *the switching condition* $(x(t_i), x(t_{i+1})) \in S_{q_i, q_{i+1}}$ *holds if*

$$i = 1, ..., r.$$

The vector $\mathcal{R} := (q_1, ...q_{r+1})^T$ *is a discrete trajectory of the hybrid control system* $\mathcal{HS}$.

Definition 10 describe the dynamic of a hybrid control system $\mathcal{HS}$. Since $x(\cdot)$ is an absolutely continuous function, Definition 10 describe a class of hybrid systems without impulse components of the continuous trajectories. Therefore, the corresponding switching sets $S_{q,q'}$ (and $S_{q_i,q_{i+1}}$) are defined for $x = x'$ (for $x(t_i) = x(t_{i+1})$).

Under the above assumptions for the given family of vector fields F, for each admissible control $u(\cdot) \in \mathcal{U}$ and for every interval $[t_{i-1}, t_i]$ (for every location $q_i \in \mathcal{R}$) there exists a unique absolutely continuous solution of the corresponding differential equation. This means that for each $u(\cdot) \in \mathcal{U}$ we have a unique absolute continuous trajectory of $\mathcal{HS}$. Moreover, the switching times $\{t_i\}$ and the discrete trajectory $\mathcal{R}$ for a hybrid control system $\mathcal{HS}$ are also uniquely defined. Therefore, it is reasonable to introduce the following concept.

Definition 11 *Let* $\mathcal{HS}$ *be a hybrid control system as defined above. For an admissible control* $u(\cdot) \in \mathcal{U}$, *the triplet* $X^u := (\tau, x(\cdot), \mathcal{R})$, *where* τ *is the set of the corresponding switching times* $\{t_i\}$, $x(\cdot)$ *and* $\mathcal{R}$ *are the corresponding continuous and discrete trajectories, is called hybrid trajectory of* $\mathcal{HS}$.

Let $\phi : \mathbb{R}^n \to \mathbb{R}$ be a continuously differentiable function. Given a hybrid control system $\mathcal{HS}$ we denote the following Mayer-type hybrid optimal control problem by OCP:

> minimize $\phi(x(1))$
> subject to $\dot{x}(t) = f_{q_i}(t, x(t), u(t))$ a.e. on$[t_{i-1}, t_i]$ $\qquad$ (6.1)
> $i = 1, ..., r + 1,\ x(0) = x_0 \in M_{q_1},\ u(\cdot) \in \mathcal{U}.$

Evidently, (6.1) is the problem of minimizing the Mayer cost functional $J(X) := \phi(x(1))$ over all hybrid trajectories X of $\mathcal{H}S$. Note that we study OCP (6.1) in the absence of target (endpoint) and state constraints. For necessary optimality conditions for (6.1) in the form of a Hybrid Maximum Principle we refer to [73, 196, 32].

Consider a hybrid control system $\mathcal{H}S$. For an admissible control function $u(\cdot) \in \mathcal{U}$ we obtain the corresponding hybrid trajectory X^u. For every interval $[t_{i-1}, t_i]$ from τ we can define the characteristic function of $[t_{i-1}, t_i]$

$$\beta_{[t_{i-1}, t_i)}(t) = \begin{cases} 1 & \text{if } t \in [t_{i-1}, t_i) \\ 0 & \text{otherwise.} \end{cases}$$

Using the introduced characteristic functions, we rewrite the state differential equations from Definition 10 for the continuous trajectory $x(\cdot)$ in the following form

$$\dot{x}(t) = \sum_{i=1}^{r+1} \beta_{[t_{i-1}, t_i)}(t) f_{q_i}(t, x(t), u(t)), \tag{6.2}$$

where $x(0) = x_0$. Under the above assumptions for the family of vector fields F, the right-hand side of the obtained differential equation (6.2) satisfies the conditions of the extended Caratheodory Theorem (see e.g., [99]). Therefore, there exists a unique (absolutely continuous) solution of (6.2). In the case of the hybrid control system $\mathcal{H}S$ we have

$$X = \mathbb{W}_n^{1,\infty}([0, 1]), \ Y = \mathbb{L}_m^2([0, 1]).$$

By $\mathbb{W}_n^{1,\infty}([0, 1])$ we denote here the Sobolev space of all absolutely continuous functions with essentially bounded derivatives. Let us introduce the operator

$$P : \mathbb{W}_n^{1,\infty}([0, 1]) \times \mathbb{L}_m^2([0, 1]) \to \mathbb{W}_n^{1,\infty}([0, 1]) \times \mathbb{R}^n,$$

where

$$P(x(\cdot), u(\cdot))\Big|_t := \begin{pmatrix} \dot{x}(t) - \sum_{i=1}^{r+1} \beta_{[t_{i-1}, t_i)}(t) f_{q_i}(t, x(t), u(t)) \\ x(0) - x_0 \end{pmatrix}.$$

Consider a regular OCP (6.1) and introduce the *Hamiltonian*

$$H(t, x, u, p) = \langle p, \sum_{i=1}^{r+1} \beta_{[t_{i-1}, t_i)}(t) f_{q_i}(t, x, u) \rangle.$$

where $p \in \mathbb{R}^n$. Since every admissible control $u(\cdot)$ determines a unique hybrid trajectory $\mathcal{X}^u$, the following cost functional

$$\tilde{J} : \mathcal{U} \to \mathbb{R}$$

such that $\tilde{J}(u(\cdot)) := J(\mathcal{X}^u)$ is well-defined. The differentiability of the given function ϕ implies the differentiability of $\tilde{J}$. The corresponding derivative is denoted by $\nabla \tilde{J}$. In the particular case of OCP (6.1) the evaluation of the adjoint operator $\nabla \tilde{J}^*$ to $\nabla \tilde{J}$ is relatively easy. We now present our main result.

Theorem 34 *Consider a regular OCP (6.1). The gradient $\nabla \tilde{J}^*(u(\cdot))$ can be found by solving the equations*

$$\dot{x}(t) = H_p(t, x(t), u(t), p(t)), \; x(0) = x_0,$$
$$\dot{p}(t) = -H_x(t, x(t), u(t), p(t)), \; p(1) = -\phi_x(x(1)), \quad\quad (6.3)$$
$$\nabla \tilde{J}^*(u(\cdot))(t) = -H_u(t, x(t), u(t), p(t)),$$

where $p(\cdot)$ is an absolutely continuous function (an "adjoint variable").

We now give give a proof-idea for the presented result. The Lagrangian of the regular problem (6.1) can be written as

$$\mathcal{L}(x(\cdot), u(\cdot), \hat{p}, p(\cdot)) = \phi(x(1)) + \langle \hat{p}, x(0) - x_0 \rangle +$$
$$+ \langle p(t), \dot{x}(t) - \sum_{i=1}^{r+1} \beta_{[t_{i-1}, t_i)}(t) f_{q_i}(t, x(t), u(t)) \rangle dt,$$

where the adjoint variable here contains two components $\hat{p} \in \mathbb{R}^n$ and $p(\cdot)$. If we differentiate the Lagrange function with respect to the adjoint variable, then we obtain the first equation from (6.3)

$$\dot{x}(t) = \sum_{i=1}^{r+1} \beta_{[t_{i-1}, t_i)}(t) f_{q_i}(t, x(t), u(t)) = H_p(t, x(t), u(t), p(t)),$$

with $x(0) = x_0$. Consider the term

$$\int_0^1 \langle p(t), \dot{x}(t) \rangle dt.$$

From the integration by part we have

$$\int_0^1 \langle p(t), \dot{x}(t) \rangle dt = \langle p(1), x(1) \rangle - \langle p(0), x(0) \rangle -$$

$$- \int_0^1 \langle \dot{p}(t), x(t) \rangle.$$

Hence

$$\mathcal{L}(x(\cdot), u(\cdot), \hat{p}, p(\cdot)) = \phi(x(1)) + \langle p(1), x(1) \rangle +$$

$$+ \langle \hat{p} - p(0), x(0) \rangle - \langle \hat{p}, x_0 \rangle - \int_0^1 \langle \dot{p}(t), x(t) \rangle dt +$$

$$+ \int_0^1 \langle p(t), \sum_{i=1}^{r+1} \beta_{[t_{i-1}, t_i)}(t) f_{q_i}(t, x(t), u(t)) \rangle dt. \tag{6.4}$$

If we differentiate $\mathcal{L}$ in (6.4) with respect to $x(\cdot)$, we can compute $\mathcal{L}_x$, $\mathcal{L}_x^*$. Thus we obtain the second relation in (6.3). Using (6.4), we also write

$$\mathcal{L}_u(x(\cdot), u(\cdot), \hat{p}, p(\cdot))v(\cdot) = - \int_0^1 H_u(t, x(t), u(t))v(t) dt$$

for every $v(\cdot) \in \mathbb{L}_m^2([0, 1])$. We obtain the last relation in (6.3)

$$\nabla \tilde{J}^*(u(\cdot))(t) = \mathcal{L}_u^*(x(\cdot), u(\cdot), \hat{p}, p(\cdot)) = -H_u(t, x(t), u(t)).$$

To make a step forward in the study of the given OCP we will discuss the necessary optimality conditions for (6.1) and some related numerical aspects. Let us formulate an easy consequence of Theorem 34.

Theorem 35 *Assume that OCP (6.1) has an optimal solution*

$$(u^{opt}(\cdot), x^{opt}(\cdot))$$

such that $u^{opt}(t) \in \text{int}\{U\}$, where $\text{int}\{U\}$ is the interior of the set U. Then $(u^{opt}(\cdot), x^{opt}(\cdot))$ can be found by solving the following equations

$$\dot{x}^{opt}(t) = H_p(t, x^{opt}(t), u^{opt}(t), p(t)),$$
$$x(0) = x_0,$$
$$\dot{p}(t) = -H_x(t, x^{opt}(t), u^{opt}(t), p(t)), \tag{6.5}$$
$$p(1) = -\phi_x(x^{opt}(1)),$$
$$H_u(t, x^{opt}(t), u^{opt}(t), p(t)) = 0.$$

Clearly, the conditions (6.5) from Theorem 35 present a necessary optimality conditions for a special case of problem (6.1). Note that the last
equation in (6.2) is consistent with the usual optimality condition

$$\nabla \tilde{J}^*(u(\cdot))(t) = 0, \ t \in [0, 1]$$

if the optimal control takes values in an open bounded control set.
Alternatively, Theorem 34 provides a basis for a wide class of the gradient-
based optimization algorithms for (6.1). We now assume that the control
set U has a so-called box-form, namely,

$$U := \{u \in \mathbb{R}^m \ : \ b_-^j \le u_j \le b_+^j, \ j = 1, ..., m\},$$

where $b_-^j, b_+^j, j = 1, ..., m$ are constants. Let us consider, for example,
the standard gradient algorithm in $\mathbb{L}_m^2([0, 1])$ (see e.g., [100, 173])

$$u^{k+1}(t) = u^k(t) - \gamma_k \nabla \tilde{J}^*(u^k(\cdot))(t), \ t \in [0, 1]$$
$$b_-^j \le u_j^{k+1}(t) \le b_+^j, \ j = 1, ..., m, \ k = 0, 1, ... \qquad (6.6)$$
$$u^0(\cdot) \in \mathcal{U},$$

where γ_k is a step-size of the gradient algorithm and

$$\{u^k(\cdot)\} \subset \mathbb{L}_m^2([0, 1])$$

is the sequence of iterations. Note that in general cases an admissible
iterative control $u^{k+1}(\cdot)$ can also be obtained by a projection

$$u^{k+1}(t) = P_U(u^k(t) - \gamma_k \nabla \tilde{J}^*(u^k(\cdot))(t)).$$

Here P_U is a projection operator on the control set U. For the Projected
Gradient Algorithm and for convergence properties of (6.6) and of some
related gradient-type optimization procedures see e.g., [173, 100].
Let us now present an implementable computational scheme that follows from our consideration presented above.

1) Choose an admissible initial control $u^0(\cdot) \in \mathcal{U}$ and the corresponding continuous trajectory $x^0(\cdot)$. Set $k = 0$.

2) Given a $x_{q_i}^k(\cdot)$ define $q_{i+1}^k, i = 1, ..., r + 1$ and

$$t_{i+1}^k := \min\{t \in [0, 1] \ : \ x_{q_i}^k(t) \bigcap S_{q_i, q_{i+1}} \ne \emptyset\}.$$

3) For the determined hybrid trajectory

$$\chi^{u^k} = (\tau^k, x^k(\cdot), \bar{q}^k)$$

solve the above equations (6.3) and define the gradient

$$\nabla \tilde{J}^*(u^k(\cdot))(t) \ \forall t \in [0, 1]$$

of the cost functional.

4) Using $\nabla \tilde{J}^*(u^k(\cdot))$, compute the iteration $u^{k+1}(t)$. Increase k by one and go to Step (2).

Note that the switching times, number of switches and switching sets in the given OCP (6.1) are assumed to be unknown. Using the iterative structure of the proposed algorithm, one can compute the corresponding approximations of the optimal trajectory, optimal switching times and optimal switching sets. The convergence result for Algorithm is presented in [31].

In this paper, we have proposed a numerical approach to a class of hybrid optimal control problems of the Mayer type. This approach is based on explicit formulae for the reduced gradient of the cost functional of the given hybrid optimal control problem. The corresponding relations make it possible to formulate first-order necessary optimality conditions for the considered hybrid optimal control problems and provide a basis for effective computational algorithms. The idea of reduced gradients can also be used for some linearization procedures of the initial optimal control problem. Note that linearization techniques have been recognized for a long time as a powerful tool for solving optimization problems. The approach proposed in this paper can be extended to some other classes of hybrid optimal control problems. Finally, note that it seems to be possible to derive necessary ϵ-optimality conditions (the ϵ-Hybrid Maximum Principle).

6.2 Optimal Control in Mechanics

In this section, we study a class of controlled mechanical systems governed by the second-order Euler-Lagrange equations or Hamilton equations. It is well-known that a large class of mechanical and physical systems admits, at least partially, a representation by these equations, which lie at the heart of the theoretical framework of physics. The important examples of controlled mechanical systems are mechanical and electromechanical plants such as diverse mechanisms, transport systems, robots, and so on [165].

The aim of our investigations is to use the variational structure of the solution to the two-point boundary-value problem for the controllable Euler-Lagrange or Hamilton equation and to propose a new computational algorithm for optimal control problems in mechanics. We consider an optimal control problem in mechanics in the general nonlinear formulation and reduce the initial optimal control problem to an auxiliary multiobjective optimization problem with constraints. This optimization problem provided a basis for solving the original optimal control problem.

The basic inspiration for modeling systems in analytical mechanics is the following variational problem

$$\text{minimize} \int_0^1 \tilde{L}(t, q(t), \dot{q}(t)) dt \tag{6.7}$$
$$\text{subject to } q(0) = c_0, \ q(1) = c_1,$$

where $\tilde{L}$ is the Lagrangian function of the (noncontrolled) mechanical system and $q(\cdot)$ is a continuously differentiable function,

$$q(t) \in \mathbb{R}^n.$$

We consider a mechanical system with n degrees of freedom, locally represented by n generalized configuration coordinates $q_1, ..., q_n$. The components $\dot{q}_\lambda(t)$, $\lambda = 1, ..., n$ of $\dot{q}(t)$ are so-called generalized velocities. We assume that the function $\tilde{L}(t, \cdot, \cdot)$ is a twice continuously differentiable function. It is also assumed that the function $\tilde{L}(t, q, \cdot)$ is a strongly convex function. The necessary conditions for the variational problem (6.7) describe the equations of motion for many mechanical systems, which are free from external influence, for appropriate choice of the Lagrangian function $\tilde{L}$. This necessary conditions are the second-order Euler-Lagrange equations [1, 9]

$$\frac{d}{dt} \frac{\partial \tilde{L}(t, q, \dot{q})}{\partial \dot{q}_\lambda} - \frac{\partial \tilde{L}(t, q, \dot{q})}{\partial q_\lambda} = 0, \ \lambda = 1, ..., n, \tag{6.8}$$
$$q(0) = c_0, \ q(1) = c_1.$$

The principle of Hamilton (see e.g., [1, 9]) gives a variational description of the solution of the two-point boundary-value problem for the Euler-Lagrange equations (6.8).

For a controlled mechanical system of n degrees of freedom with a Lagrangian $L(t, q, \dot{q}, u)$ we introduce the equations of motion

$$\frac{d}{dt} \frac{\partial L(t, q, \dot{q}, u)}{\partial \dot{q}_\lambda} - \frac{\partial L(t, q, \dot{q}, u)}{\partial q_\lambda} = 0,$$

$$q(0) = c_0, \quad q(1) = c_1,$$
(6.9)

where $u(\cdot) \in \mathcal{U}$ is a control function from the set of admissible controls $\mathcal{U}$. Let

$$\mathcal{U} := \{v(\cdot) \in \mathbb{L}_m^2([0, 1]) \ : \ v(t) \in U \text{ a.e. on } [0, 1]\}$$

and $U := \{u \in \mathbb{R}^m \ : \ b_{1,\nu} \leq u_\nu \leq b_{2,\nu}, \ \nu = 1, ..., m\}$, where $b_{1,\nu}, b_{2,\nu}, \nu = 1, ..., m$ are constants. The introduced set $\mathcal{U}$ provides a standard example of an admissible control set (see e.g., [120]). In specific cases we consider the following set of admissible controls $\mathcal{U} \cap \mathbb{C}_m^1(0, 1)$. We also examine the given controlled mechanical system in the absence of external forces. The Lagrangian function L depends directly on the control function $u(\cdot)$. We assume that the function $L(t, \cdot, \cdot, u)$ is a twice continuously differentiable function and $L(t, q, \dot{q}, \cdot)$ is a continuously differentiable function. For a fixed admissible control $u(\cdot) \in \mathcal{U}$ we obtain the usual (noncontrolled) mechanical system with $\tilde{L}(t, q, \dot{q}) \equiv L(t, q, \dot{q}, u(t))$ and the corresponding Euler-Lagrange equation (6.8). It is assumed that the function $L(t, q, \cdot, u)$ is a strongly convex function, i.e., for any $(t, q, \dot{q}, u) \in \mathbb{R} \times \mathbb{R}^n \times \mathbb{R}^n \times \mathbb{R}^m$ and $\xi \in \mathbb{R}^n$ the following inequality

$$\sum_{\lambda,\theta=1}^n \frac{\partial^2 L(t, q, \dot{q}, u)}{\partial \dot{q}_\lambda \partial \dot{q}_\theta} \xi_\lambda \xi_\theta \geq \alpha \sum_{\lambda=1}^n \xi_\lambda^2, \ \alpha > 0$$

holds. This convexity condition is a direct consequence of the representation $\frac{1}{2} \dot{q}^T M(t, u) \dot{q}$ for the kinetic energy of a mechanical system. The matrix $M(t, u)$ here is a positive definite matrix. Under the above-mentioned assumptions for the Lagrangian function L the two-point boundary-value problem (6.9) has a solution for every control function $u(\cdot) \in \mathcal{U}$ [107]. We assume that (6.9) has a unique solution for every $u(\cdot) \in \mathcal{U}$. Given an admissible control function $u(\cdot) \in \mathcal{U}$ the solution to the boundary-value problem (6.9) is denoted by $q^u(\cdot)$. We will call (6.9) an *Euler-Lagrange control system*. Note that (6.9) is a system of second-order differential equations.

Example 11 *We consider a linear mass-spring system* [165] *attached to a moving frame. The considered control* $u(\cdot) \in \mathcal{U} \cap \mathbb{C}_1^1(0, 1)$ *is the velocity of the frame. By* ω *we denote the mass of the system. The kinetic energy* $0.5\omega(\dot{q} + u)^2$ *depends directly on* $u(\cdot)$, *and so does the Lagrangian function*

$$L(q, \dot{q}, u) = 0.5(\omega(\dot{q} + u)^2 - \kappa q^2), \ \kappa \in \mathbb{R}_+,$$

yielding the equation of motion (6.9)

$$\frac{d}{dt} \frac{\partial L(t, q, \dot{q}, u)}{\partial \dot{q}} - \frac{\partial L(t, q, \dot{q}, u)}{\partial q} = \omega(\ddot{q} + \dot{u}) + \kappa q = 0.$$

By κ *we denote here the elasticity coefficient of the system.*

Some important controlled mechanical systems have the Lagrangian function of the following form (see e.g., [165])

$$L(t, q, \dot{q}, u) = L_0(t, q, \dot{q}) + \sum_{\nu=1}^{m} q_\nu u_\nu.$$

In this special case we have

$$\frac{d}{dt} \frac{\partial L_0(t, q, \dot{q})}{\partial \dot{q}_\lambda} - \frac{\partial L_0(t, q, \dot{q})}{\partial q_\lambda} = \begin{cases} u_\lambda & \lambda = 1, ..., m, \\ 0 & \lambda = m + 1, ..., n. \end{cases}$$

and the control function $u(\cdot)$ can be interpreted as an external force. Let us now pass on to the Hamiltonian formulation. For the Euler-Lagrange control system (6.9) we introduce the generalized momenta $p_\lambda := L(t, q, \dot{q}, u)/\partial \dot{q}_\lambda$ and define the Hamiltonian function $H(t, q, p, u)$ as a Legendre transform of $L(t, q, \dot{q}, u)$, i.e.

$$H(t, q, p, u) := \sum_{\lambda=1}^{n} p_\lambda \dot{q}_\lambda - L(t, q, \dot{q}, u).$$

In the case of hyperregular Lagrangians $L(t, q, \dot{q}, u)$ (see e.g., [1]) the Legendre transform $\mathcal{L}$ is a diffeomorphism

$$\mathcal{L} : (t, q, \dot{q}, u) \rightarrow (t, q, p, u).$$

Using the introduced Hamiltonian $H(t, q, p, u)$, we can rewrite the equations of motion (6.9)

$$\dot{q}_\lambda(t) = \frac{\partial H(t, q, p, u)}{\partial p_\lambda}, \quad q(0) = c_0, \quad q(1) = c_1,$$

$$\dot{p}_\lambda(t) = -\frac{H(t, q, p, u)}{\partial q_\lambda}, \quad \lambda = 1, ..., n \,. \tag{6.10}$$

Under the above-mentioned assumptions the boundary-value problem (6.10) has a solution for every $u(\cdot) \in \mathcal{U}$. We will call (6.10) a *Hamilton control system*. A main advantage of (6.10) in comparison with (6.9) is that (6.10) immediately constitutes a control system in standard state space form [120], with state variables (q, p) (in physics usually called the *phase variables*). Consider the system of Example 11 with

$$H(q, p, u) = \frac{1}{2}\omega(\dot{q}^2 - u^2) + \frac{1}{2}\kappa q^2 = \frac{1}{2\omega}p^2 + \frac{1}{2}\kappa q^2 - up.$$

The Hamilton equations in this case are given as

$$\dot{q} = \frac{\partial H(q, p, u)}{\partial p} = \frac{1}{\omega}p - u,$$

$$\dot{p} = -\frac{\partial H(q, p, u)}{\partial q} = -\kappa q.$$

Note that if $L(t, q, \dot{q}, u)$ is given as

$$L(t, q, \dot{q}, u) = L_0(t, q, \dot{q}) + \sum_{\nu=1}^{m} q_\nu u_\nu$$

then we have

$$H(t, q, p, u) = H_0(t, q, p) - \sum_{\nu=1}^{m} q_\nu u_\nu,$$

where $H_0(t, q, p)$ is the Legendre transform of $L_0(t, q, \dot{q})$.
Let us now consider the following OCP with constraints

$$\text{minimize } J(q(\cdot), u(\cdot)) := \int_0^1 f_0(t, q(t), u(t))dt$$

$$\text{subject to (6.9), } u(t) \in U \ t \in [0, 1], \tag{6.11}$$

$$h_j(u(\cdot)) \leq 0 \ \forall j \in I,$$

$$g_k(q(\cdot))(t) \leq 0 \ \forall k \in K, \ \forall t \in [0, 1],$$

where $h_j : \mathbb{L}^2_m([0,1]) \to \mathbb{R}$, $g_k : \mathbb{C}^1_n(0,1) \to \mathbb{C}(0,1)$ for $j \in I$ and $k \in K$. Let $f_0 : [0,1] \times \mathbb{R}^n \times \mathbb{R}^m \to \mathbb{R}$ be a continuous function. By I and K we denote finite sets of index values. The ensuring analysis is restricted to the convex on $\mathbb{R}^n \times \mathbb{R}^m$ function $f_0(t,\cdot,\cdot)$ and to the proper convex on $\mathbb{L}^2_m([0,1])$ and on $\mathbb{C}^1_n(0,1)$ functionals $h_j(\cdot)$, $j \in I$ and $g_k(\cdot)(t)$, $k \in K$, $t \in [0,1]$. We assume that the boundary-value problem (6.9) has a unique solution and (6.11) has an optimal solution. The class of optimal control problems of the type (6.11) is broadly representative [120, 178]. Let $(q^{opt}(\cdot), u^{opt}(\cdot))$ be an optimal solution of (6.11). Note that we formulate the initial optimal control problem for the Euler-Lagrange control system. Clearly, it is also possible to use the Hamiltonian formulation. Note that a variety of constraints may be represented in the above form, including the initial conditions, boundary conditions and interior point conditions of the general form. For example, if the initial optimal control problem contains the target constraints

$$\hat{h}_j(q(1)) \le 0 \ \forall j \in I, \ \hat{h}_j : \mathbb{R}^n \to \mathbb{R},$$

then $h_j(u(\cdot)) := \hat{h}_j(q^u(1))$ for all $j \in I$.

We mainly focus our attention on the application of a direct numerical method to the constrained optimal control problem (6.11). A great amount of works is devoted to the direct or indirect numerical methods for optimal control problems (see [173, 204, 178, 27] and references therein). One can find a fairly complete review of the main results in [178, 20, 27].

It is common knowledge that an optimal control problem involving ordinary differential equations can be formulated in various ways as an optimization problem in a suitable function space [90, 120, 178, 20]. For example, the original problem (6.11) can be expressed as an infinite-dimensional optimization problem over the set of control functions $u(\cdot) \in \mathcal{U}$ (or $u(\cdot) \in \mathcal{U} \cap \mathbb{C}^1_m(0,1)$).

$$
\begin{aligned}
&\text{minimize } \hat{J}(u(\cdot)) \\
&\text{subject to } u(\cdot) \in \mathcal{U}, \\
&h_j(u(\cdot)) \le 0 \ \forall j \in I, \ G_k(u(\cdot))(t) \le 0 \\
&\forall k \in K, \ \forall t \in [0,1]
\end{aligned}
\tag{6.12}
$$

with the aid of the functions $\hat{J} : \mathbb{L}^2_m([0,1]) \to \mathbb{R}$ and

$$G_k : \mathbb{L}^2_m([0,1]) \to \mathbb{C}(0,1) \ \forall k \in K :$$

$$\hat{J}(u(\cdot)) := J(q^u(\cdot), u(\cdot)) = \int_0^1 f_0(t, q^u(t), u(t))dt,$$

$$G_k(u(\cdot))(t) := g_k(q^u(\cdot))(t) \ \forall k \in K, \ \forall t \in [0, 1].$$

The minimization problem (6.12) can be solved by using some numerical algorithms (e.g., by applying a first order method [101, 173]). For example, the implementation of the method of feasible directions is presented in [178].

Example 12 *Using the* Euler-Lagrange *control system of* Example 11, *we formulate the optimal control problem*

$$\text{minimize } J(q(\cdot), u(\cdot)) := -\int_0^1 (u(t) + q(t))dt$$

$$\text{subject to } \ddot{q}(t) + \frac{\kappa}{\omega}q(t) = -\dot{u}(t) \ q(0) = 0, \ q(1) = 1,$$

$$u(\cdot) \in \mathbb{C}_1^1(0, 1), \ 0 \le u(t) \le 1 \ \forall t \in [0, 1],$$

$$\dot{u}(t) \ge 0, \ \int_0^1 u(t)dt \le \frac{1}{2}, \ q(t) \le 3 \ \forall t \in [0, 1].$$

Let $\omega \ge 4\kappa/\pi^2$. The solution $q^u(\cdot)$ of the boundary-value problem is

$$q^u(t) = C^u \sin(t\sqrt{\kappa/\omega}) -$$

$$- \int_0^t \sqrt{\kappa/\omega} \sin(\sqrt{\kappa/\omega}(t - \tau))\dot{u}(\tau)d\tau,$$

where

$$C^u = \frac{1}{\sin\sqrt{\kappa/\omega}}[1 + \int_0^1 \sqrt{\kappa/\omega} \sin(\sqrt{\kappa/\omega}(t - \tau))\dot{u}(\tau)d\tau]$$

is a constant. Consequently,

$$\hat{J}(u(\cdot)) = -\int_0^1 [u(t) + q^u(t)]dt = -\int_0^1 [u(t) +$$

$$+ C^u \sin(t\sqrt{\kappa/\omega}) - \int_0^t \sqrt{\kappa/\omega} \sin(\sqrt{\kappa/\omega}(t - \tau))\dot{u}(\tau)d\tau]dt.$$

Moreover,

$$h_1(u(\cdot)) = -\dot{u}(t), \ h_2(u(\cdot)) = \int_0^1 u(t)dt - \frac{1}{2}$$

and $g_1(q(\cdot))(t) = q(t) - 1$. The above-mentioned additional conditions $\omega \geq 4\kappa/\pi^2$ and $\dot{u}(t) \geq 0$ imply

$$\sin(\sqrt{\kappa/\omega}(t - \tau)) \geq 0, \ \sin(t\sqrt{\kappa/\omega}) \geq 0,$$

$$\int_0^t \sqrt{\kappa/\omega} \sin \sqrt{\kappa/\omega}(t - \tau)\dot{u}(\tau)d\tau \geq 0,$$

$$\int_0^1 \sqrt{\kappa/\omega} \sin \sqrt{\kappa/\omega}(t - \tau)\dot{u}(\tau)d\tau \geq 0, \ C^u \geq 0.$$

We claim that $u^{opt}(t) \equiv 0.5$ is an optimal solution of the given optimal control problem. Note that this result is consistent with the Bauer *maximum principle (see e.g., [5]). For $u^{opt}(\cdot)$ we obtain the optimal trajectory*

$$q^{opt}(t) = \frac{\sin(t\sqrt{\kappa/\omega})}{\sin \sqrt{\kappa/\omega}}.$$

Evidently, we have $\sqrt{\kappa/\omega} \leq \pi/2$, $q^{opt}(t) \leq 3$, where $q^{opt}(\cdot) \in \mathbb{C}_1^1(0, 1)$.

Finally note that the class of impulsive control systems (see e.g., [59]) can also be described by the presented controllable Euler-Lagrange or Hamilton equations. Thus, one can write the constrained optimal control problems for impulsive systems in the form (6.11).

6.3   The Variational Approach

An effective numerical procedure, as a rule, uses the specific character of the concrete problem. Our aim is to consider the variational description of the optimal control problem (6.11). Let

$$\Gamma := \{\gamma(\cdot) \in \mathbb{C}_n^1([0, 1]) \ : \ \gamma(0) = c_0, \ \gamma(1) = c_1\}.$$

The following theorem is an immediate consequence of the classical Hamilton principle from analytical mechanics (see [34, 35]).

Theorem 36 *Let the Lagrangian $L(t, q, \dot{q}, u)$ be a strongly convex function of the variables $\dot{q}_i$, $i = 1, ..., n$. Assume that the boundary-value problem (6.9) has a unique solution for every*

$$u(\cdot) \in \mathcal{U} \bigcap \mathbb{C}^1_m(0, 1).$$

The function $q^u(\cdot)$ with $u(\cdot) \in \mathcal{U} \cap \mathbb{C}^1_m(0, 1)$ is the solution of the boundary-value problem (6.9) if and only if

$$q^u(\cdot) = \operatorname{argmin}_{q(\cdot) \in \Gamma} \int_0^1 L(t, q(t), \dot{q}(t), u(t))dt.$$

For a fixed admissible control function $u(\cdot)$ we introduce two following functionals

$$T(q(\cdot), z(\cdot)) := \int_0^1 [L(t, q(t), \dot{q}(t), u(t)) - L(t, z(t), \dot{z}(t), u(t))]dt,$$

$$V(q(\cdot)) := \max_{z(\cdot) \in \Gamma} \int_0^1 [L(t, q(t), \dot{q}(t), u(t)) - L(t, z(t), \dot{z}(t), u(t))]dt.$$

Let us also consider the (nonempty) set Θ of all functions $q(\cdot) \in \Gamma$ satisfying the inequalities constraints from (6.12). We now give a variational description of the admissible solutions $q^u(\cdot)$ to problem (6.9) [34].

Theorem 37 *Let the Lagrangian $L(t, q, \dot{q}, u)$ be a strongly convex function of the variables $\dot{q}_i$, $i = 1, ..., n$. Assume that the boundary-value problem (6.9) has a unique solution for every*

$$u(\cdot) \in \mathcal{U} \bigcap \mathbb{C}^1_m(0, 1).$$

The function $q^u(\cdot) \in \Theta$ is a solution of the boundary-value problem (6.9) for a control function $u(\cdot) \in \mathcal{U} \cap \mathbb{C}^1_m(0, 1)$ if and only if

$$q^u(\cdot) = \operatorname{argmin}_{q(\cdot) \in \Theta} V(q(\cdot)) \tag{6.13}$$

Let $q^u(\cdot) \in \Theta$ be a unique solution of (6.9) for an input

$$u(\cdot) \in \mathcal{U} \bigcap \mathbb{C}^1_m(0, 1).$$

Using the Hamilton principle, we obtain

$$\min_{q(\cdot)\in\Theta} V(q(\cdot)) =$$

$$= \min_{q(\cdot)\in\Theta} \max_{z(\cdot)\in\Gamma} \int_0^1 [L(t,q(t),\dot{q}(t),u(t)) - \int_0^1 L(t,z(t),\dot{z}(t),u(t))]dt =$$

$$= \min_{q(\cdot)\in\Theta} \int_0^1 L(t,q(t),\dot{q}(t),u(t))dt - \min_{z(\cdot)\in\Gamma} \int_0^1 L(t,z(t),\dot{z}(t),u(t))dt =$$

$$= \int_0^1 L(t,q^u(t),\dot{q}^u(t),u(t))dt-$$

$$- \int_0^1 L(t,q^u(t),\dot{q}^u(t),u(t))dt = V(q^u(\cdot)) = 0.$$

If the condition (6.13) holds, then $q^u(\cdot)$ is a solution of the boundary-value problem (6.9). This completes the proof.

The presented theorems make it possible to express the initial optimal control problem (6.11) as a multiobjective optimization problem over the set of admissible control functions and generalized coordinates

$$\text{minimize } J(q(\cdot),u(\cdot)) \text{ and } P(q(\cdot))$$
$$\text{subject to } (q(\cdot),u(\cdot)) \in \Gamma \times (\mathcal{U} \bigcap \mathbb{C}^1_m(0,1)),$$
$$h_j(u(\cdot)) \leq 0 \,\forall j \in I, \; g_k(q(\cdot))(t) \leq 0$$
$$\forall k \in K, \; \forall t \in [0,1], \qquad (6.14)$$

or

$$\text{minimize } J(q(\cdot),u(\cdot)) \text{ and } V(q(\cdot))$$
$$\text{subject to } (q(\cdot),u(\cdot)) \in \Gamma \times (\mathcal{U} \bigcap \mathbb{C}^1_m(0,1)),$$
$$h_j(u(\cdot)) \leq 0 \,\forall j \in I, \; g_k(q(\cdot))(t) \leq 0$$
$$\forall k \in K, \; \forall t \in [0,1], \qquad (6.15)$$

where

$$P(q(\cdot)) := \int_0^1 L(t,q(t),\dot{q}(t),u^{opt}(t))dt.$$

We define the objective functionals $P(\cdot)$ and $V(\cdot)$ for an optimal control function $u(\cdot) = u^{opt}(\cdot)$. The auxiliary minimizing problems (6.14) and (6.15) are multiobjective optimization problems (see e.g., [193]).

The set $\Gamma \times (\mathcal{U} \cap \mathbb{C}_m^1(0,1))$ is a convex set. Since $f_0(t, \cdot, \cdot)$, $t \in [0,1]$ is a convex function, $J(q(\cdot), u(\cdot))$ is convex. If $P(\cdot)$ (or $V(\cdot)$) is a convex functional, then we deal with a convex multiobjective minimization problem (6.14) (or (6.15)).

The variational description of the solution of the two-point boundary-value problem for the Lagrange equations (6.9) eliminates the differential equations from consideration. The problems (6.14) and (6.15) provide a basis for numerical algorithms to the initial optimal control problem (6.11). The auxiliary optimization problem (6.14) has two objective functionals. For (6.14) we introduce the Lagrange function [80]

$$\Lambda(t, q(\cdot), u(\cdot), \mu, r, s, l) := \mu_1 J(q(\cdot), u(\cdot)) + \mu_2 P(q(\cdot)) +$$
$$+ \sum_{j \in I} r_j h_j(u(\cdot)) + \sum_{k \in K} \bar{s}_k(g_k(q(\cdot))) +$$
$$+ l|(\mu, r, \bar{s})|\text{dist}_{\Gamma \times (\mathcal{U} \cap \mathbb{C}_m^1(0,1))}\{(q(\cdot), u(\cdot))\},$$

where $\bar{s}_k$ is a continuous linear functional from the (topological) dual space to $\mathbb{C}(0,1)$ and $\text{dist}_{\Gamma \times (\mathcal{U} \cap \mathbb{C}_m^1(0,1))}\{\cdot\}$ denotes the distance function

$$\text{dist}_{\Gamma \times (\mathcal{U} \cap \mathbb{C}_m^1(0,1))}\{(q(\cdot), u(\cdot))\} := \inf\{\|(q(\cdot), u(\cdot)) -$$
$$- \varrho\|_{\mathbb{C}_n^1(0,1) \times \mathbb{C}_m^1(0,1)}, \ \varrho \in \Gamma \times (\mathcal{U} \cap \mathbb{C}_m^1(0,1))\}$$

associated with $\Gamma \times (\mathcal{U} \cap \mathbb{C}_m^1(0,1))$. We use the following notation

$$\mu := (\mu_1, \mu_2)^T \in \mathbb{R}_+^2, \ r := (r_j)^T \in \mathbb{R}^I$$

and $\bar{s} := (\bar{s}_k)^T$, $k \in K$. Recall that a feasible point $(q^*(\cdot), u^*(\cdot))$ is called *Pareto optimal* for the multiobjective problem (6.15) if there is no feasible point $(q(\cdot), u(\cdot))$ for which

$$J(q(\cdot), u(\cdot)) < J(q^*(\cdot), u^*(\cdot)) \text{ and } P(q(\cdot)) < P(q^*(\cdot)).$$

A necessary condition for $(q^*(\cdot), u^*(\cdot))$ to be a Pareto optimal solution to (6.15) in the sense of Kuhn-Tucker (see [80, 193]) is that for every $l \in \mathbb{R}$ sufficiently large there exist $\mu^* > 0$, $r^* \geq 0$ and $\bar{s}^* \neq 0$ such that

$$\sum_{j \in I} r_j^* h_j(u^*(\cdot)) + \sum_{k \in K} \bar{s}_k^*(g_k(q^*(\cdot))) = 0,$$
$$0 \in \partial_{(q(\cdot), u(\cdot))} \Lambda(t, q^*(\cdot), u^*(\cdot), \mu^*, r^*, \bar{s}^*, l). \tag{6.16}$$

By $\partial_{(q(\cdot), u(\cdot))}$ we denote here the *generalized gradient* of the Lagrange function Λ [80]. If $P(\cdot)$ is a convex functional, then the necessary condition (6.16) is also sufficient for $(q^*(\cdot), u^*(\cdot))$ to be a Pareto optimal solution to (6.15). Let $\aleph$ be a set of all Pareto optimal solutions $(q^*(\cdot), u^*(\cdot))$ to (6.14). Since $(q^{opt}(\cdot) u^{opt}(\cdot)) \in \aleph$, the above conditions (6.16) are satisfied also for this optimal pair $(q^{opt}(\cdot) u^{opt}(\cdot))$. Note that one can investigate the auxiliary minimization problem (6.15) in an similar way.

A direct implementation of the necessary conditions (6.16) is often not practical. Using a discretization of (6.14) we can obtain a finite-dimensional approximating problem. Note that discrete approximation techniques have been recognized as a powerful tool for solving optimal control problems [178, 20]. Let N be a sufficiently large positive integer number and

$$\mathcal{G}_N := \{t_0 = 0, t_1, ..., t_N = 1\}$$

be a (possible nonequidistant) partition of $[0, 1]$ with

$$\max_{0 \leq i \leq N-1} |t_{i+1} - t_i| \leq \xi_N.$$

We assume that $\lim_{N \to \infty} \xi_N = 0$. Define

$$\Delta t_{i+1} := t_{i+1} - t_i, \quad i = 0, ..., N - 1$$

and consider the following finite-dimensional optimization problem

$$
\begin{aligned}
&\text{minimize } J_N(q_N(\cdot), u_N(\cdot)) \text{ and } P_N(q_N(\cdot)), \\
&\text{subject to } q_N(t_0) = c_0, \ q_N(t_N) = c_1, \\
&b_1 \leq u_N(t_i) \leq b_2, \ h_j(u_N(\cdot)) \leq 0 \ \forall j \in I, \\
&g_k(q_N(\cdot))(t_i) \leq 0 \ \forall k \in K, \forall t_i \in \mathcal{G}_N,
\end{aligned}
\tag{6.17}
$$

where b_1 and b_2 are constant vectors,

$$J_N(q_N(\cdot), u_N(\cdot)) := \sum_{i=0}^{N-1} f_0(t_i, q^i, u^i) \Delta t_{i+1},$$

$$P_N(q_N(\cdot)) := \sum_{i=0}^{N-1} L(t_i, q^i, \dot{q}_N(t_i), u^{opt}(t_i)) \Delta t_{i+1}$$

and

$$q_N(t) := \sum_{i=0}^{N-1} \phi_i(t) q^i, \; q^i = q(t_i),$$

$$u_N(t) := \sum_{i=0}^{N-1} \phi_i(t) u^i, \; u^i = u(t_i), \; t \in [0,1], \; t_i \in \mathcal{G}_N,$$

$$\phi_i(t) := \begin{cases} 1 & \text{if } t \in [t_i, t_{i+1}[, \; i = 0, ..., N-1; \\ 0 & \text{otherwise.} \end{cases}$$

In effect, we deal with the Euclidean spaces $\mathbb{L}_n^{2,N}(\mathcal{G}_N)$ and $\mathbb{L}_m^{2,N}(\mathcal{G}_N)$ of the piecewise constant trajectories $q_N(\cdot)$ and piecewise constant control functions $u_N(\cdot)$. As we can see the Banach space $\mathbb{C}_n^1(0,1)$ and the Hilbert space $\mathbb{L}_m^2([0,1])$ are replaced by $\mathbb{L}_n^{2,N}(\mathcal{G}_N)$ and by $\mathbb{L}_m^{2,N}(\mathcal{G}_N)$, respectively. The discrete optimization problem (6.17) approximates the infinite dimensional optimization problem (6.14). We assume that the set of all Pareto optimal solution of the discrete problem (6.17) is nonempty. If $P(\cdot)$ is a convex functional, then the discrete multiobjective optimization problem (6.17) is also a convex problem. Let

$$\Gamma_N := \{\gamma_N(\cdot) \in \mathbb{L}_n^{2,N}(\mathcal{G}_N) \; : \; \gamma_N(t_0) = c_0, \; \gamma_N(t_N) = c_1\},$$
$$\mathcal{U}_N := \{u_N(\cdot) \in \mathbb{L}_m^{2,N}(\mathcal{G}_N) \; : \; b_1 \leq u_N(t_i) \leq b_2, \; i \in \mathcal{G}_N\}$$

For (6.17) we also can introduce the Lagrange function Λ_N (see [193])

$$\Lambda_N(t_i, q_N, u_N, \mu, r, s, \sigma) := \mu_1 J_N(q_N(\cdot), u_N(\cdot)) +$$
$$+ \mu_2 P_N(q_N(\cdot)) + \sum_{j \in I} r_j h_j(u_N(\cdot)) +$$
$$+ \sum_{k \in K} s_k(t_i) g_k(q_N(\cdot))(t_i) + \langle \sigma_1(t_i), b_1 - u_N(t_i) \rangle_{\mathbb{R}^m} +$$
$$+ \langle \sigma_2(t_i), u_N(t_i) - b_2 \rangle_{\mathbb{R}^m},$$

where $\mu, r, s(\cdot), \sigma(\cdot)$ are Lagrange multipliers such that

$$s(t_i) := (s_k(t_i))^T, \; k \in K, \; \sigma(t_i) := (\sigma_1(t_i), \sigma_2(t_i))^T$$

and $\sigma_1(t_i), \sigma_2(t_i) \in \mathbb{R}^m$. We now consider the corresponding necessary (Kuhn-Tucker) conditions for $(q_N^*(\cdot), u_N^*(\cdot))$ to be a Pareto optimal solution to (6.17). In this case we have the following Kuhn-Tucker system

$$
\mu_1^* \begin{pmatrix} \nabla_{q_N(\cdot)} J_N(q_N^*(\cdot), u_N^*(\cdot)) \\ \nabla_{u_N(\cdot)} J_N(q_N^*(\cdot), u_N^*(\cdot)) \end{pmatrix} + \mu_2^* \begin{pmatrix} \nabla_{q_N(\cdot)} P_N(q_N^*(\cdot)) \\ 0 \end{pmatrix} +
$$

$$
+ \begin{pmatrix} 0 \\ \sum_{j \in I} r_j^* \nabla_{u_N(\cdot)} h_j(u_N^*(\cdot)) \end{pmatrix} + \begin{pmatrix} \sum_{k \in K} s_k^*(t_i) \nabla_{q_N(\cdot)} g_k(q_N^*(\cdot))(t_i) \\ 0 \end{pmatrix} +
$$

$$
+ \begin{pmatrix} 0 \\ \langle \sigma_1^*(t_i), -e \rangle_{\mathbb{R}^m} + \langle \sigma_2^*(t_i), e \rangle_{\mathbb{R}^m} \end{pmatrix} = 0, \tag{6.18}
$$

$$
\sum_{j \in I} r_j^* h_j(u_N^*(\cdot)) + \sum_{k \in K} s_k^*(t_i) g_k(q_N^*(\cdot))(t_i) +
$$

$$
+ \langle \sigma_1^*(t_i), b_1 - u_N^*(t_i) \rangle_{\mathbb{R}^m} + \langle \sigma_2^*(t_i), u_N^*(t_i) - b_2 \rangle_{\mathbb{R}^m} = 0,
$$

$$
q_N^*(t_0) - c_0 = 0, \quad q_N^*(t_N) - c_1 = 0,
$$

$$
\mu^* > 0, \quad r^* \geq 0, \quad s^* \neq 0, \quad \sigma^* \neq 0,
$$

where $\nabla_{q_N(\cdot)}$, $\nabla_{u_N(\cdot)}$ stand for partial derivatives, μ^*, r^*, $s^*(\cdot)$ and $\sigma^*(\cdot)$ are the (Pareto) optimal Lagrange multipliers [193]. By $e \in \mathbb{R}^m$ we denote a unit vector. If $P(\cdot)$ is a convex functional, then the necessary condition (6.18) is also sufficient for $(q_N^*(\cdot), u_N^*(\cdot))$ to be a Pareto optimal solution to (6.17). An optimal solution $(q_N^{opt}(\cdot), u_N^{opt}(\cdot))$ to the finite-dimensional problem (6.17) belongs to the set of all Pareto optimal solutions of (6.17). Thus $(q_N^{opt}(\cdot), u_N^{opt}(\cdot))$ satisfies the presented conditions (6.18). In a similar manner, one can derive the Kuhn-Tucker conditions for a finite-dimensional optimization problem over the set of variables (q^i, u^i), $i = 0, ..., N$.
The necessary optimality conditions (6.18) reduce the finite dimensional multiobjective optimization problem to a problem of finding a zero of nonlinear functions. Such a problem can be solved by using some gradient-based or Newton-like methods [101, 173]. From the viewpoint of numerical mathematics we solve the optimal control problem in mechanics approximately. We now propose a (conceptual) computational algorithm based on the finite-dimensional approximations (6.17) and on the corresponding Kuhn-Tucker system (6.18).
Fix a small parameter $\epsilon > 0$.

- 1. Choose the initial control $u^{(0)}(\cdot) \in \mathcal{U} \cap \mathbb{C}_m^1(0, 1)$ which satisfies

$$
h_j(u^{(0)}(\cdot)) \leq 0 \ \forall j \in I,
$$

$$
g_k(q^{(0)}(\cdot))(t) \leq 0,
$$

where $q^{(0)}(\cdot)$ is a solution of (6.9) for $u^{(0)}(\cdot)$. Define

$$u_N^{(0)}(t_i) := u^{(0)}(t_i), \quad q_N^{(0)}(t_i) := q^{(0)}(t_i),$$

where $t_i \in \mathcal{G}_N$. Set $a = 0$.

- 2. Compute

$$P_N = (q_N(\cdot)) \sum_{i=0}^{N-1} L(t_i, q^i, \dot{q}_N(t_i), u^{(a)}(t_i)) \Delta t_{i+1}.$$

Increase a by one.

- 3. Solve the following Kuhn-Tucker system of algebraic equations and inequalities

$$\mu_1 \begin{pmatrix} \nabla_{q_N(\cdot)} J_N(q_N(\cdot), u_N(\cdot)) \\ \nabla_{u_N(\cdot)} J_N(q_N(\cdot), u_N(\cdot)) \end{pmatrix} + \mu_2 \begin{pmatrix} \nabla_{q_N(\cdot)} P_N(q_N(\cdot)) \\ 0 \end{pmatrix} +$$

$$+ \begin{pmatrix} 0 \\ \sum_{j \in I} r_j \nabla_{u_N(\cdot)} h_j(u_N(\cdot)) \end{pmatrix} + \begin{pmatrix} \sum_{k \in K} s_k(t_i) \nabla_{q_N(\cdot)} g_k(q_N(\cdot))(t_i) \\ 0 \end{pmatrix} +$$

$$+ \begin{pmatrix} 0 \\ -\sigma_1(t_i) + \sigma_2(t_i) \end{pmatrix} = 0,$$

$$\sum_{j \in I} r_j h_j(u_N(\cdot)) + \sum_{k \in K} s_k(t_i) g_k(q_N(\cdot))(t_i) +$$

$$+ \langle \sigma_1(t_i), b_1 - u_N(t_i) \rangle_{\mathbb{R}^m} + \langle \sigma_2(t_i), u_N(t_i) - b_2 \rangle_{\mathbb{R}^m} = 0,$$

$$q_N(t_0) - c_0 = 0, \quad q_N(t_N) - c_1 = 0,$$

$$\mu > 0, \quad r \geq 0, \quad s \neq 0, \quad \sigma \neq 0.$$

Let $(q_N^{(a)}(\cdot), u_N^{(a)}(\cdot))$ be a solution of this system. If

$$\|u_N^{(a)} - u_N^{(a-1)}\| \leq \epsilon,$$

then STOP.

- 4. Otherwise, go to the Step 2.

For the aims of solving the Kuhn-Tucker system one can use, for example, a variant of the Newton-type method. Note that the similar approach can also be considered for the auxiliary problem (6.15). We refer to [34, 35] for corresponding convergence result. We now apply the computational algorithm given above to a simple "mechanical" OCP.

Example 13 *Consider the optimal control problem from* Example 12. *We put* $\omega = 1$, $\kappa = 1$. *Evidently, the condition* $\omega \geq 4\kappa/\pi^2$ *holds. We use the following initial control function* $u^{(0)}(t) = t$, $t \in [0,1]$. *It can easily be shown that* $u^{(0)}(\cdot)$ *is an admissible control function. We use an equidistant partition* $\mathcal{G}_N$ *of* $[0,1]$ *and* $\Delta t_i = \Delta t = 1/N$ *for all* $i = 1,...,N$. *We now apply the proposed computational scheme for* $N = 100$, $\epsilon = 10^{-3}$ *and consider the approximating finite-dimensional optimization problem as an optimization problem over the set of variables* (q^i, u^i), $i = 0,...,N$. *In fact, we deal with the following problem*

$$\text{minimize } -\sum_{i=0}^{N-1}(q^i + u^i)$$

$$\text{and } \frac{1}{2}[\sum_{i=0}^{N-1}(N(q^{i+1} - q^i) + u_N^{opt}(t_i))^2 - (q^i)^2]$$

$$\text{subject to } q^0 = 0, \; q^N = 1,$$

$$-u^i \leq 0, \; u^i - 1 \leq 0, \; i = 0,...,N;$$

$$u^i - u^{i+1} \leq 0, \; i = 0,...,N-1; \; \sum_{i=0}^{N-1}u^i - \frac{1}{2} \leq 0,$$

$$q^i - 3 \leq 0, \; i = 1,...,N-1.$$

Note that we approximate the derivatives $\dot{q}$ *and* $\dot{u}$ *by* $(q^{i+1} - q^i)/\Delta t$ *and* $(u^{i+1} - u^i)/\Delta t$, *respectively. The* Lagrange *function* Λ_N *in this case can be written as follows*

$$\Lambda_N(\mu_1, \mu_2, q^0, ..., q^N, u^0, ..., u^N, r_1, r_2, s, \sigma_1, \sigma_2) =$$

$$= -\mu_1 \sum_{i=0}^{N-1}(q^i + u^i) +$$

$$+ \mu_2 \frac{1}{2}[\sum_{i=0}^{N-1}(N(q^{i+1} - q^i) + u_N^{opt}(t_i))^2 - (q^i)^2] +$$

$$+ \sum_{i=0}^{N-1}r_1^i(u^i - u^{i+1}) + r_2(\sum_{i=0}^{N-1}u^i - \frac{1}{2}) +$$

$$+ \sum_{i=1}^{N-1}s^i(q^i - 3) - \sum_{i=0}^{N}\sigma_1^i u^i + \sum_{i=0}^{N}\sigma_2^i(u^i - 1),$$

where $\mu_1, \mu_2, r_2 \in \mathbb{R}$, $r_1 \in \mathbb{R}^N$, $s, \sigma_1, \sigma_2 \in \mathbb{R}^{N+1}$. We now put $\mu_1 = \mu_2 = 1$. For some positive parameters r_1, r_2, s, σ_1 and σ_2 we apply the Newton-Raphson *method (see e.g., [101]) to the corresponding* Kuhn-Tucker *system for problem under consideration. The above implementation of Conceptual Algorithm was carried out, using the standard* MATLAB *packages. The computed optimal control $u_N(\cdot)$ and the computed optimal trajectory $q_N(\cdot)$ have the properties*

$$\max_{0 \le i \le N} |u_N(t_i) - u^{opt}(t_i)| \le 10^{-3},$$

$$\max_{0 \le i \le N} |q_N(t_i) - q^{opt}(t_i)| \le 2 \cdot 10^{-3}.$$

The computed optimal objective value is (-1.0442). Note that the exact optimal objective value in this example is

$$J(q^{opt}(\cdot), u^{opt}(\cdot)) \approx -1.0463.$$

The implementation of the presented algorithm requires a first approximation $u^{(0)}(\cdot)$ to an optimal control $u^{opt}(\cdot)$. The efficiency of this algorithm essentially depends on $u^{(0)}(\cdot)$.

In many real-life problems, especially in mechanical, electrical and chemical engineering, the dynamics of the system can be complex. In most cases the problems are nonlinear and contain some additional constraints of the general nature. These factors lead to several difficulties, among which the problem of multiple local minima has become a new focus of the practical computational optimization. The methods most often employed in the literature to address this problem use *control parametrization*. This approach also converts the infinite dimensional problem into a finite one in which the optimization is performed over a set of parameters. The methods used for the discretization vary from simple piecewise constant functions to complicated polynomials and finite elements. Note that the control vector parametrization is often combined with the optimization techniques based on the *orthogonal collocation method*. Different approaches (for example, sequential and simultaneous deterministic techniques, stochastic approaches) have been taken into account for the dynamic nature of the formulation (see e.g., [110, 147, 208, 93] and references therein).

It is evident that methods of control vector parametrization and orthogonal collocations can also be applied to the initial optimal control problems of the type (6.11) or directly to the auxiliary multiobjective optimization problems (6.14) and (6.15). In this case as well as by using the proposed algorithm the global minimum can be obtained by enumeration of the possible minima. However, in contrast to the proposed algorithm the convergence of the above parametrization-based numerical approach is guaranteed only for a very limited class of problems (see e.g., [93]). Drawbacks to the methods of control parametrization and orthogonal collocations include a large increase in the size of the variable space and the number of constraints. Also, the type of discretization used for the state and control profiles can have a dramatic effect on the solution due to the error introduced in the approximations. On the other hand, the speed and accuracy of the proposed algorithm are determined by a concrete solution procedure to the Kuhn-Tucker system (6.18). For instance, in Example 13 we use a second-order Newton-based method. On this basis, one can draw inferences about the convergence properties of the considered numerical algorithm and corresponding approximation errors.

Since the variational structure of the optimal control problem (6.11) is independent of the possible state/control constraints, the proposed theoretical approach as well as the computational scheme from this section can be extended to some classes of problems with general constraints (for example, for problems with DAE's). In this case one need to choose a suitable discretization procedure for the optimal control problem under consideration and modify the Kuhn-Tucker system (6.11). Finally note that it seems to be possible to apply the above theoretical and computational schemes to some classes of hybrid optimal control problems in mechanics.

Chapter 7

Regularity Conditions for OCPs

In the preceding chapters we have considered regular OCPs. Under the regularity assumptions we have, for one, deduced the gradient formulae for the abstract OCP, for the initial OCPs of the type (1.2) and (1.2a) and for the discretized problems. Generally the regularity conditions and the so called *constraint qualifications* play an important role in the numerical analysis of OCPs. This chapter is devoted to a class of OCPs in Banach spaces involving equality constraints with *stable operators*. We now give a short introduction to the theory of stable operators, present the new analytical properties of stable and expanding operators obtained by the author and propose the new constraint qualifications for the above abstract problems.

7.1 Stable Operators

We deal with operator equations

$$A(x) = a, \tag{7.1}$$

where X, Y are real Banach spaces and $A : X \to Y$, is, in general, a nonlinear operator and $x \in X$, $a \in Y$. The symbol $A : X \to Y$ will mean a single valued mapping whose domain of definition is X and whose range is contained in Y, that is , for every $x \in X$ the mapping A assigns a unique element $A(x) \in Y$. We now start recalling some basic concepts and preliminary results (see e.g., [168, 169, 221, 222, 223, 224] and [25]).

Definition 12 *An operator $A : X \to Y$ is called stable if there is a strictly monotone increasing and continuous function $g : \mathbb{R}_+ \to \mathbb{R}_+$ with*

$$g(0) = 0, \ \lim_{t \to +\infty} g(t) = +\infty.$$

such that

$$\|A(x_1) - A(x_2)\|_Y \geq g(\|x_1 - x_2\|_X) \ \forall x_1, x_2 \in X.$$

The function $g(\cdot)$ is called a stabilizing function of the operator A.

Let $\{H, \langle \cdot \rangle_H\}$ be a real Hilbert space. By $\langle \cdot \rangle_H$ we denote the inner product of H. We now define strongly stable operators in a real Hilbert space.

Definition 13 *An operator $B : H \to H$ is called strongly stable if there is a number $c > 0$ such that*

$$|\langle B(h_1) - B(h_2), h_1 - h_2 \rangle_H| \geq c \|h_1 - h_2\|_H^2 \ \forall h_1, h_2 \in H.$$

Let $A : X \to Y$ be stable. Then, for each $a \in Y$, the operator equation (7.1) has at most one solution. Moreover, we have the continuous dependence of the solution on the right-hand side of the equation $A(x) = a$ (see [25]). From Definition 12 it follows that the solution $\hat{x}$ of (7.1) is stable in the following sense: for each $\epsilon > 0$ there exists a number $\delta(\epsilon) > 0$ such that

$$\|a_1 - a_2\|_Y < \delta(\epsilon),$$

where $a_1, a_2 \in R(A)$ always implies that $\|\hat{x}_1 - \hat{x}_2\|_X < \epsilon$ for the corresponding solution $\hat{x}_1, \hat{x}_2 \in X$ of the problems $A(x) = a_1$ and $A(x) = a_2$, respectively. We now suggest the following definition.

Definition 14 *An operator $A : X \to Y$ is called expanding if there is a number $d > 0$ such that*

$$\|A(x_1) - A(x_2)\|_Y \geq d \|x_1 - x_2\|_X \ \forall x_1, x_2 \in X.$$

Definition 15 *We call a stable operator A hyper-stable, if there exists a strictly monotone increasing and continuous function*

$$\tilde{g} : \mathbb{R}_+ \to \mathbb{R}_+$$

with

$$\tilde{g}(0) = 0, \ \lim_{t \to +\infty} \tilde{g}(t) = +\infty$$

such that the stabilizing function $g(\cdot)$ of A satisfies the inequality

$$k g(t/k) \geq \tilde{g}(t) \ \forall t, k \in \mathbb{R}_+.$$

It is evident that an expanding operator A is a stable operator with the stabilizing function $g(t) = d \cdot t$, $t \geq 0$. We refer to the book of the author [25] for some concrete examples of stable and expanding operators in Banach and Hilbert spaces.

The stable and strongly stable operators play an important role in the general theory of discretization methods. In the book of the author [25] we present some basic results of this theory in connection with the stable operators. The aim of our book is to study the class of differentiable stable operators and to prove a general solvability theorem for nonlinear operator equations with differentiable stable operators. In parallel with the given nonlinear equation (7.1) we examine the corresponding linearization and discretizations of (7.1) and also obtain solvability results. In connection with stable operators we discuss the well-posedness concepts of Hadamard [114], Tykhonov [211] and some related topics. In [25] we also study *local stability* concepts, *local ill-posedness* and some aspects of approximate solutions (*minimum norm solution* and *quasi-solution*). Note that the mathematical theory of well-posedness is comprehensively presented in the literature (see e.g., [87, 127, 213, 164, 225]). As possible applications of the general theory of stable operators we consider in [25] not only operator equations but also some classes of optimization problems, namely, quadratic minimization problems and optimal control problems with constraints. We examine an existence result for a quadratic minimization problem and present constrain qualifications for abstract minimization and abstract optimal control problems. Moreover, we study some theoretical aspects of optimization problems.

Let us discuss some properties of the introduced operators. We refer to [25] for full proofs of the presented results.

Lemma 3 *Let $A : X \to Y$ be hyper-stable and* Fréchet *differentiable at $x_0 \in X$. Then the linear operator $A'(x_0) \in L(X, Y)$ is stable.*

Lemma 4 *Let A be an expanding and* Fréchet *differentiable at a point $x_0 \in X$*

$$\|A(x_1) - A(x_2)\|_Y \geq d\|x_1 - x_2\|_X \ \forall x_1, x_2 \in X.$$

Then $A'(x_0)$ is expanding with the same constant $d > 0$, that means

$$\|A'(x_0)x_1 - A'(x_0)x_2\|_Y \geq d\|x_1 - x_2\|_X \ \forall x_1, x_2 \in X.$$

Lemma 5 *Let* $A : X \to Y$ *be stable and continuous. Then the range* $R(A)$ *is a closed subset of* Y.

Lemma 6 *Let* $A : X \to Y$ *be hyper-stable and* Fréchet *differentiable at* $x_0 \in X$. *Then the operator*

$$(A'(x_0))^{-1} : R(A'(x_0)) \subseteq Y \to X$$

is linear and continuous.

Lemma 7 *Let* $\{A^s\}$, $s \in \mathbb{N}$ *be a sequence of stable continuous operators*

$$A^s : X \to Y,$$

and $\{g^s(\cdot)\}$ *be a sequence of stabilizing functions conforming to* $\{A^s\}$. *Assume that*

$$\inf\{g^s(\cdot)\} \geq g(\cdot),$$

where $g : \mathbb{R}_+ \to \mathbb{R}_+$ *is a strictly monotone increasing and continuous function with*

$$g(0) = 0, \ \lim_{t \to +\infty} g(t) = +\infty,$$

and

$$R(A^s) = Y, \ \forall s \in \mathbb{N}.$$

If the operator $A : X \to Y$ *is the uniform limit of* $\{A^s\}$, *then* A *is a stable continuous operator and* $R(A) = Y$. *The function* $g(\cdot)$ *is a stabilizing function of* A.

In [25] we also study the theory of linear stable operators.

Lemma 8 *Let* $\{\mathcal{A}^s\}$ $s \in \mathbb{N}$ *be a sequence of stable linear continuous operators* $\mathcal{A} \in L(X, Y)$ *and* $\{g^s(\cdot)\}$ *be a sequence of stabilizing functions conforming to* $\{\mathcal{A}^s\}$. *Assume that*

$$\inf\{g^s(\cdot)\} \geq g(\cdot),$$

where $g : \mathbb{R}_+ \to \mathbb{R}_+$ *is a strictly monotone increasing and continuous function with*

$$g(0) = 0, \ \lim_{t \to +\infty} g(t) = +\infty.$$

Let

$$R(\mathcal{A}^s) = Y, \ \forall s \in \mathbb{N}.$$

If $\lim_{s \to \infty} \|\mathcal{A}^s - \mathcal{A}\|_{L(X,Y)} = 0$, *then* $\mathcal{A} \in L(X, Y)$ *is a stable operator and* $R(\mathcal{A}) = Y$. *The function* $g(\cdot)$ *is a stabilizing function for* $\mathcal{A}$.

In the special case $Y = X$ we introduce the set $\beth(X, X) \subset L(X, X)$ of stable linear continuous surjective operators $\mathcal{A} : X \to X$ such that

$$g(t) \geq \tilde{g}(t) \ \forall t \in \mathbb{R}_+,$$

where $g(\cdot)$ is a stabilizing function of $\mathcal{A}$ and $\tilde{g} : \mathbb{R}_+ \to \mathbb{R}_+$ is a strictly monotone increasing and continuous function with

$$\tilde{g}(0) = 0, \ \lim_{t \to +\infty} \tilde{g}(t) = +\infty.$$

One can prove the following theorem.

Theorem 38 *The set $\beth(X, X)$ is a* Banach *algebra with unity element (the identity operator $I : X \to X$).*

If we deal with the set $\daleth(X, Y) \subset L(X, Y)$ of all expanding linear continuous surjective operators $\mathcal{A}$ which have an inverse linear continuous operator $\mathcal{A}^{-1}$, then we have our next result [25].

Theorem 39 *Let $\mathcal{A} \in \daleth(X, Y)$, $\|\mathcal{A}x_1 - \mathcal{A}x_2\|_Y \geq d\|x_1 - x_2\|_X$ and*

$$\mathcal{D} \in L(X, Y)$$

such that

$$\|\mathcal{D}\|_{L(X,Y)} < \frac{1}{\|\mathcal{A}^{-1}\|_{L(Y,X)}} \text{ and } \|\mathcal{D}\|_{L(X,Y)} < \tilde{d} < d.$$

Then the operator $(\mathcal{A} + \mathcal{D})$ is expanding and invertible. Moreover, $(\mathcal{A} + \mathcal{D})^{-1}$ is a linear continuous operator, i.e.,

$$(\mathcal{A} + \mathcal{D}) \in \daleth(X, Y).$$

In the book [25] we consider the relations between stable operators and some standard classes of operators in Banach spaces, namely, compact operators, symmetric operators in Hilbert spaces and accretive and monotone operators. We now present some corresponding results.

Theorem 40 *Let $A : X \to Y$ be expanding and* Fréchet *differentiable at $x_0 \in X$. If*

$$\dim R(A'(x_0)) = \infty,$$

then $A'(x_0)$ is a noncompact operator.

Theorem 41 *A number $\lambda \in \mathbb{R}$ is a regular value of a symmetric operator*

$$\mathcal{B} \in L(H, H)$$

if and only if $(\lambda I - \mathcal{B})$ is an expanding operator

$$\|(\lambda I - \mathcal{B})h\|_H \geq d\|h\|_H \ \forall h \in H.$$

Let also $\|\mathrm{Res}(\lambda; \mathcal{B})\|_{L(H,H)} = r < 1$. Then the operator $\mathcal{B}$ is expanding.

Theorem 42 *Let $\mathcal{B} \in L(H, H)$ be a positive symmetric operator. Then $\lambda = -1$ is a regular value of the operator $\mathcal{B}$. If $\mathcal{B}$ is surjective, then this operator is maximal monotone.*

Theorem 43 *Let $B : H \to H$ be a monotone and continuously* Fréchet *differentiable operator on the real* Hilbert *space H. Let*

$$R(B'(\tilde{h})) = H$$

for a vector $\tilde{h} \in H$. Then B is a maximal monotone and maximal accretive operator.

As can be readily appreciated, for expanding operators $\mathcal{B} \in L(H, H)$ with the property $\dim R(\mathcal{B}) = \infty$ the well known Fredholm Alternative (see e.g., [181]) does not hold.

Definition 16 *Let $\mathcal{A} \in L(X, Y)$. The linear problem*

$$\mathcal{A}x = a, \ x \in X, \ a \in Y \tag{7.2}$$

is called Hadamard *well-posed if the continuous linear operator $\mathcal{A}$ is bijective and the inverse operator $\mathcal{A}^{-1}$ is continuous, i.e.,*

$$\mathcal{A}^{-1} \in L(Y, X).$$

Otherwise, the linear problem is called ill-posed in the Hadamard *sense.*

The linear problem (7.2) can be considered as a particular case of the general problem (7.1). Let $A : X \rightarrow Y$. We now require for A the existence of Fréchet derivatives $A'(x_0)$ at the point $x_0 \in X$. It makes sense to compare the nonlinear equation (7.1) with its linearization $A'(x_0)x = a$ at the point $x_0 \in X$. Note that one can distinguish three types of Hadamard ill-posedness (see [134]):

- $\mathcal{A}$ is not surjective,

- $\mathcal{A}$ is not injective,

- the inverse operator $\mathcal{A}^{-1}$ is not continuous.

Using the above results, we now formulate two main theorems of the above-mentioned author's book [25].

Theorem 44 *Let X, Y be real* Banach *spaces. Let*

- *an operator $A : X \rightarrow Y$ be expanding and continuously* Fréchet *differentiable at every point $x \in X$;*

- $R(A'(\tilde{x})) = Y$ *for a $\tilde{x} \in X$.*

Then the linear problem

$$A'(x_0)x = a, \ x_0, x \in X, \ a \in Y \tag{7.3}$$

is Hadamard *well-posed for every $x_0 \in X$.*

Theorem 45 *Let X, Y be real* Banach *spaces and let*

- $A : X \rightarrow Y$ *be an expanding and continuously* Fréchet *differentiable operator;*

- $R(A'(\tilde{x})) = Y$ *for an element $\tilde{x} \in X$.*

Then the given operator A has an inverse $A^{-1} : Y \rightarrow X$ and the equation (7.1) has a unique solution for every $a \in Y$.

It is evident that Theorem 44 and Theorem 45 represent a solvability result for the class of operator equations (7.1) and linearized equations (7.3) with expanding, continuously Fréchet differentiable operators. Theorem 44 defines a class of Hadamard well-posed linear problems. In effect the surjectivity of the Fréchet derivative at an arbitrary point of an expanding operator involves the Hadamard well-posedness of all linear problems of the type (7.3). Before closing this section we give two corollaries of these main theorems.

Corollary 1 *Let X, Y be real* Banach *spaces, $A : X \to Y$ be continuously* Fréchet *differentiable at every point $x \in X$ and the mapping $A'(x)$ be expanding. If*

$$R(A'(\tilde{x})) = Y$$

for a $\tilde{x} \in X$, then the adjoint operators $(A'(x))^* \in L(Y^*, X^*)$ are expanding for every $x \in X$.

The next result is in fact a new fixed point theorem for some classes of expanding operators in Banach spaces [29].

Corollary 2 *Let X be real* Banach *spaces and let*

- *$A : X \to X$ be expanding with the expansivity constant $d > 1$ and continuously* Fréchet *differentiable operator;*

- *$R(A'(\tilde{x})) = X$ for a $\tilde{x} \in X$.*

Then the mapping A has a unique fixed point.

For other corollaries of Theorem 44 and Theorem 45 see [25, 29].

7.2   Constraint Qualifications for OCPs

Consider a smooth optimization problem with equality constraints given in operator form

$$\begin{aligned}
&\text{minimize } \psi_0(\xi) \\
&\text{subject to } \xi \in Q := \{\xi \in X \ : \ A(\xi) = a\},
\end{aligned} \tag{7.4}$$

where $\psi_0 : X \to \mathbb{R}$ is a sufficiently smooth function in a real Banach space X and $A : X \to Y$ is an operator between real Banach spaces X and Y. We assume A to be Fréchet differentiable on X. For (7.4) we introduce the Lagrange function

$$\mathcal{L}(\lambda_0, \lambda, \xi) = \lambda_0 \psi_0(\xi) + \langle \lambda, A(\xi) - a \rangle_Y$$

where $\lambda_0 \in \mathbb{R}$ and λ is a linear functional from the dual space Y^*. Here we have

$$\langle \lambda, \cdot \rangle_Y : Y \to \mathbb{R}.$$

Let ξ^{opt} be a *local solution* of (7.4). We are interested in the case when the equality constraints are *regular*.

Definition 17 *Let a mapping $A : X \to Y$ be Fréchet differentiable at some point $\xi \in X$. We say the mapping A is regular at ξ if*

$$R(A'(\xi)) = Y.$$

The mapping A is called *nonregular* (*irregular, abnormal, degenerate*) if the regularity condition is not satisfied. Various regularity conditions are comprehensively discussed by Mangasarian and Fromowitz [151], Robinson [182], Zowe and Kurcyusz [226], Auslender [15], Malanowski [149], and in the books of Ioffe and Tichomirov [120], Bertsekas [45], Fletcher [101], and others. We now formulate necessary optimality conditions for the problem (7.4) in the form of the generalized Lagrange multiplier rule (for the analytical background see [120, 46] or [124]).

Proposition 4 *Let x^{opt} be a local solution to (7.4). Let the objective functional ψ_0 and the constraints mapping A be Fréchet differentiable at ξ^{opt}. Assume that the range $R(A'(\xi^{opt}))$ of the mapping*

$$\xi \to A'(\xi^{opt})\xi$$

is closed. Then there exist Lagrange *multipliers*

$$\lambda_0^{opt} \geq 0, \ \lambda^{opt} \in Y^*$$

not all equal to zero such that

$$\mathcal{L}'_\xi(\lambda_0^{opt}, \lambda^{opt}, \xi^{opt}) = \lambda_0^{opt} \psi_0'(\xi) + \langle \lambda^{opt}, A'(\xi^{opt}) \rangle_Y = 0.$$

If, moreover, A is regular at x^{opt} and the mapping $A'(\xi^{opt})$ is continuous, then $\lambda_0^{opt} \neq 0$.

Clearly, in the case $\lambda_0 \neq 0$ one can put $\lambda_0 = 1$. Various supplementary conditions have been proposed under which it is possible to assert that Lagrange multiplier rule holds with $\lambda_0 = 1$. These conditions are called *constraint qualifications* (or *regularity conditions*) (see e.g., [80, 124]). To mention just a few: Mangasarian-Fromowitz constraint qualifications [56], Kurcyusz-Robinson-Zowe regularity conditions [226] and Ioffe regularity conditions [120].

Note that in the case of a convex optimization problem (2.1) we have known Slater conditions [124]. If the Fréchet derivative $A'(\xi^{opt})$ is not onto, then the Lagrange necessary conditions for optimality are trivially satisfied with $\lambda_0 = 0$ and provide no additive information about solutions of (7.4). Note that the development of optimality conditions for nonregular problems (7.4) has also become an active research topic (see e.g., [138, 122, 60] and references therein).

We now apply our surjectivity result, namely, Theorem 44 to the regularity problem (see also [25]).

Theorem 46 *Let $A : X \to Y$ be expanding and continuously* Fréchet *differentiable on a real* Banach *space X. If $A'(\tilde{\xi})$ is regular at an arbitrary point $\tilde{\xi} \in X$, then the mapping $A'(\xi)$ is regular at every point $\xi \in X$.*

Since the range $R(A'(\xi))$ of the expanding continuous mapping $A'(\xi)$ is a closed set, the Lagrange optimality conditions are satisfied. Moreover, A is regular at ξ^{opt} and $A'(\xi^{opt})$ is continuous. It is important to keep in mind that Theorem 46 makes it possible to verify the surjectivity of the operator $A'(\xi^{opt})$ by using an arbitrary admissible point $\xi \in Q$! Under assumptions of Theorem 46, the regularity of the operator $A'(\xi^{opt})$ at the optimal solution ξ^{opt} follows from the regularity of $A'(\xi)$ at a point $\xi \in Q$.

In [25] we also briefly consider a nonregular case. The main idea of a constructive approach to the Lagrange-type multiplier rule in this case is based on a construction proposed in [60]. One replaces the operator $A'(\xi^{opt})$, which is non onto, with a linear operator $\Psi_p(\xi^{opt})$, related to the p-th Taylor polynomial of A at ξ^{opt}, which is onto. Let

$$A : X \to Y$$

be a p-times continuously Fréchet differentiable operator. Moreover, we assume that the space Y is decomposed into a direct sum

$$Y = Y_1 \oplus ... \oplus Y_p,$$

where

$$Y_1 = \mathrm{cl}\{R(A'(\xi))\},$$
$$Y_i = \mathrm{cl}\{\mathrm{Sp}(R(P_{Z_i}A^{(i)}(\xi))(\cdot)^i)\},\ i = 2, ..., p-1,$$
$$Y_p = Z_p,$$

and Z_i is a closed complementary subspace for $(Y_1 \oplus ... \oplus Y_{i-1})$ with respect to Y, $i = 2, ..., p$. Here $P_{Z_i} : Y \to Z_i$ is the projection operator onto Z_i along $(Y_1 \oplus ... \oplus Y_{i-1})$ with respect to Y, $i = 2, ..., p$. Define the mappings [123]

$$\psi_i : X \to Y_i,\ \psi_i(\xi) = P_{Y_i}A(\xi),$$

where $P_{Y_i} : Y \to Y_i$ is the projection operator onto Y_i along

$$(Y_1 \oplus ... \oplus Y_{i-1} \oplus Y_{i+1} \oplus ... \oplus Y_p)$$

with respect to Y, $i = 1, ..., p$.

Definition 18 *The linear operator* $\Psi_p(\xi_0)(\xi) \in L(X, Y)$, $\xi, \xi_0 \in X$,

$$\Psi_p(\xi_0)(x) = \psi_1'(\xi_0) + \frac{1}{2}\psi_2''(\xi_0)(\xi) + ... + \frac{1}{p!}\psi_p^{(p)}(\xi_0)(\xi)^{p-1},$$

is called a p-factor-operator. We say the mapping A is p-regular at $\xi_0 \in X$ *along an element* $x \in X$ *if* $R(\Psi_p(\xi_0)(\xi)) = Y$.

Consider the optimization problem (7.4) with a nonregular operator A. We define the *p-factor-Lagrange function*

$$\mathcal{L}^p(\lambda_0(\xi), \lambda(\xi), \xi_0, \xi) = \lambda_0(\xi)\psi_0(\xi_0) + \sum_{i=1}^{p}\langle\lambda_i(\xi), \psi_i^{(i-1)}(\xi_0)(\xi)^{i-1}\rangle_Y,$$

where $\xi, \xi_0 \in X$, $\lambda_0(\xi) \in \mathbb{R}$ and $\lambda_i(\xi) \in Y_i^*$, $i = 1, ..., p$. The function $\mathcal{L}^p$ is a generalization of the Lagrange function $\mathcal{L}$ and it reduces to the Lagrange function for the regular case. Using $\mathcal{L}^p$, one can prove the following necessary optimality conditions (see [60]).

Proposition 5 *Let X and $Y = Y_1 \oplus ... \oplus Y_p$ be real* Banach *spaces, $\psi_0 :$ $X \to \mathbb{R}$ a twice continuously* Fréchet *differentiable function , and let $A : X \to Y$ be a p-times continuously* Fréchet *differentiable mapping. Assume that for an element*

$$\xi \in \bigcap_{i=1}^{p} \mathrm{Ker}\{\psi_i^{(i)}(\xi^{opt})\}$$

the set $R(\Psi_p(\xi^{opt})(\xi))$ is closed in Y. If ξ^{opt} is a local solution to problem (7.4), then there exist multipliers $\lambda_0^{opt}(\xi) \in \mathbb{R}$ and

$$\lambda_i^{opt}(\xi) \in Y_i^*, \ i = 1, ..., p,$$

such that they do not all vanish, and

$$\mathcal{L}_\xi'^p(\lambda_0^{opt}(\xi), \lambda^{opt}(\xi), \xi^{opt}, \xi) = \lambda_0^{opt}(\xi)\psi_0(\xi^{opt}) +$$

$$+ \sum_{i=1}^{p} \langle (\psi_i^{(i)}(\xi^{opt})(\xi)^{i-1})^* \lambda_i^{opt}(\xi) \rangle_{Y^*} = 0.$$

If, moreover, $R(\Psi_p(\xi^{opt})(\xi)) = Y$, then $\lambda_0^{opt}(\xi) \neq 0$.

We are ready to formulate our p-regularity condition for the differentiable expanding operators [25].

Theorem 47 *Let $A : X \to Y$ be expanding and continuously* Fréchet *differentiable on a real* Banach *space X. If A is p-regular at an arbitrary point $\xi_0 \in X$ along an element $\xi \in X$, then the mapping A is p-regular at every point of X along the element $x \in X$. In particular,*

$$R(\Psi_p(\xi^{opt})(\xi)) = Y.$$

We now consider the generalization of (7.4) which involves a control variable u along with the state variable x

$$\begin{aligned} &\text{minimize } J(x, u) \\ &\text{subject to } F(x, u) = 0_Z, \\ &(x, u) \in S, \end{aligned} \qquad (7.5)$$

where X, Y are real Banach spaces, $J : X \times Y \to R$ is an objective functional and

$$F : X \times Y \to Z$$

is a given mapping. Here Z is a real Banach space. By S we denote a nonempty subset of $X \times Y$. In [25] we study the abstract optimal control problems (7.5) with an expanding mapping F.

Definition 19 *The mapping $F(\cdot, \cdot)$ is called expanding if there is a number $d > 0$ such that*

$$\|F(x_1, u_1) - F(x_2, u_2)\|_Z \geq d(\|x_1 - x_2\|_X + \|u_1 - u_2\|_Y)$$

for all $(x_1, u_1), (x_2, u_2) \in X \times Y$.

Note that we use the standard norm $\| \cdot \|_X + \| \cdot \|_Y$ in the product space $X \times Y$. Let us introduce the Lagrangian of problem (7.5)

$$\mathcal{L}(\lambda_0, \lambda, x, u) := \lambda_0 J(x, u) + \langle \lambda, F(x, u) \rangle_Z,$$

where $\lambda_0 \in \mathbb{R}$ and $\lambda \in Z^*$. Let F and J be Fréchet differentiable. By F' and J' we denote the corresponding Fréchet derivatives with respect to (x, u). The next theorem generalizes the known Lagrange multiplier rule.

Proposition 6 *Let (x^{opt}, u^{opt}) be a local solution of problem (7.5). Let the functional $J(\cdot, \cdot)$ be Fréchet differentiable at (x^{opt}, u^{opt}) and the mapping $F(\cdot, \cdot)$ be Fréchet differentiable in a neighborhood of (x^{opt}, u^{opt}). Suppose that $F'(\cdot, \cdot)$ is continuous at (x^{opt}, u^{opt}) and the range*

$$R(F'(x^{opt}, u^{opt}))$$

is a closed set. Then there are a real number $\lambda_0^{opt} \geq 0$ and a continuous linear functional $\lambda^{opt} \in Z^*$ with $(\lambda_0^{opt}, \lambda^{opt}) \neq (0, 0_{Z^*})$ and

$$\mathcal{L}'(\lambda_0^{opt}, \lambda^{opt}, x^{opt}, u^{opt}) = (\lambda_0^{opt} J'(x^{opt}, u^{opt}) +$$
$$+ \langle \lambda^{opt}, F'(x^{opt}, u^{opt}) \rangle_Z)(x - x^{opt}, u - u^{opt}) \geq 0,$$

for all $(x, u) \in S$. If, in addition to the assumptions given above, some regularity condition is fulfilled, then $\lambda_0^{opt} > 0$.

If the mapping $F'(x^{opt}, u^{opt})$ is surjective, then we have the regular case, namely, $\lambda_0^{opt} > 0$. Evidently, the regular case is of prime interest for applications. Note that the regularity conditions are also frequently used in proofs of convergence of some numerical schemes for optimal control problems [154, 178]. Using the main Theorem 44 we can prove the following result from [25].

Theorem 48 *Let X, Y, Z be real* Banach *spaces. Let $F : X \times Y \to Z$ be expanding and continuously* Fréchet *differentiable on $X \times Y$. If $F'(\tilde{x}, \tilde{u})$ is regular at an arbitrary point $(\tilde{x}, \tilde{u}) \in X \times Y$, then the mapping $F'(x, u)$ is regular at every point $(x, u) \in X \times Y$.*

Some optimality criteria for the abstract optimal control problem (7.5) contain the regularity condition of the mapping $x \to F(x, u^{opt})$ at the point x^{opt} [120]. Now we assume that the mapping

$$F(\cdot, u^{opt}) : X \to Z$$

is expanding and continuously Fréchet differentiable on X with

$$R(F'_x(\tilde{x}, u^{opt})) = Z, \ \tilde{x} \in X,$$

where F'_x is the partial derivative of F with respect to x. In a similar way, we can deduce the regularity of the mapping $F(\cdot, u^{opt})$ at the optimal point x^{opt}.

Example 14 *In [25] we apply the presented results to the following OCP*

$$\text{minimize } J(x(\cdot), u(\cdot)) = \int_0^1 f_0(t, x(t), u(t))dt$$

subject to $\dot{x}(t) = f(t, x(t), u(t))$ a.e. on $[0, 1]$,

$x(0) = x_0$,

$u(t) \in U$ a.e. on $[0, 1]$,   (7.6)

$g(x(t)) = 0$.

In addition to the usual assumptions we suppose that $f(t, \cdot, \cdot)$ is an expanding function for all $t \in [0, 1]$ and that $g(\cdot)$ is a differentiable expanding function. Common encountered regularity conditions for optimal control problems of the type (7.6) are based on the controllability or *local controllability assumptions for the linearized equality constraint [124]. In practice, the controllability or local controllability conditions are difficult to verify. Moreover, the conditions of the standard criterion use the unknown information about the optimal pair (x^{opt}, u^{opt}). Using Theorem 48, we can deduce the regularity condition for some special cases of OCPs [25]. The original problem (7.6) can be expressed as the infinite-dimensional minimization problem*

$$\text{minimize } \tilde{J}(u(\cdot))$$
$$\text{subject to } u(\cdot) \in \mathcal{U}, \ \tilde{g}(u(\cdot))(t) = 0, \ \forall t \in [0, 1],$$

with the aid of functions

$$\tilde{J} : \mathbb{L}_m^2([0, 1]) \to \mathbb{R}, \ \tilde{g} : \mathbb{L}_m^2([0, 1]) \to \mathbb{C}^1([0, 1], \mathbb{R}) :$$

$$\tilde{J}(u(\cdot)) := \int_0^1 f_0(t, x^u(t), u(t))dt, \ \tilde{g}(u(\cdot))(t) := g(x^u(t)) \ \forall t \in [0, 1].$$

The function $g(\cdot)$ is continuously differentiable. Evidently, we have

$$\|g(u_1(\cdot))(\cdot) - g(u_2(\cdot))(\cdot)\|_{\mathbb{C}^1([0,1],\mathbb{R})} \geq$$
$$\geq \max_{t \in [0,1]} \|g(x^{u_1}(t)) - g(x^{u_2}(t))\| \geq d_1 \max_{t \in [0,1]} \|x^{u_1}(t) - x^{u_2}(t)\| =$$
$$= d_1 \max_{t \in [0,1]} \| \int_0^t f(\tau, x^{u_1}(\tau), u_1(\tau)) - f(\tau, x^{u_2}(\tau), u_2(\tau))d\tau\|.$$

where $d_1 > 0$. The linear operator

$$\rho(\cdot) \to \int_0^t \rho(\tau)d\tau,$$

where $\rho(\cdot) \in \mathbb{C}([0, 1], \mathbb{R}^n)$, is a linear homeomorphism (see [5]). Therefore

$$d_1 \max_{t \in [0,1]} \| \int_0^t f(\tau, x^{u_1}(\tau), u_1(\tau)) - f(\tau, x^{u_2}(\tau), u_2(\tau))d\tau\| \geq$$
$$\geq d_2 \max_{t \in [0,1]} \|f(t, x^{u_1}(t), u_1(t)) - f(t, x^{u_2}(t), u_2(t))\|.$$

for a positive number d_2. Since $f(t, \cdot, \cdot)$ is an expanding function for all $t \in [0, 1]$, we obtain

$$d_2 \max_{t \in [0,1]} \|f(t, x^{u_1}(t), u_1(t)) - f(t, x^{u_2}(t), u_2(t))\| \geq$$
$$\geq d_3 \max_{t \in [0,1]} (\|x^{u_1}(t) - x^{u_2}(t)\| + \|u_1(t) - u_2(t)\|) \geq$$
$$\geq d_3 \max_{t \in [0,1]} \|u_1(t) - u_2(t)\| \geq d_3 \|u_1(\cdot) - u(\cdot)\|_{\mathbb{L}_m^2([0,1])}$$

where $d_3 > 0$. Hence

$$\|g(u_1(\cdot))(\cdot) - g(u_2(\cdot))(\cdot)\|_{\mathbb{C}^1([0,1],\mathbb{R})} \geq d\|u_1(\cdot) - u(\cdot)\|_{\mathbb{L}^2_m([0,1])}$$
$$\forall u_1(\cdot), u_2(\cdot) \in \mathbb{L}^2_m([0,1]),$$

where $d > 0$. Under the assumptions given above, the function $\tilde{g}(\cdot)$ is Fréchet *differentiable* [125, 126]. *In specific cases the* Fréchet *derivative $\tilde{g}'(u(\cdot))$ of $\tilde{g}(\cdot)$ can be computed explicitly* [178, 123]. *If*

$$R(\tilde{g}'(u(\cdot))) = \mathbb{C}^1([0,1], \mathbb{R})$$

for a control function $u(\cdot) \in \mathbb{L}^2_m([0,1])$, then the operator $\tilde{g}$ is regular at every $u(\cdot) \in \mathbb{L}^2_m([0,1])$ and satisfies all conditions of Theorem 48.

Finally note, that in connection with the classic controllability results from the control theory, the question arises which conditions are necessary or sufficient for

$$R(\mathcal{A}) = Y,$$

where $\mathcal{A} : X \rightarrow Y$ is a linear operator between two real Banach spaces. We refer to [143] for the basic theoretical facts. It seems to be possible to obtain some new controllability conditions for control systems (1.3) by using our main analytical results from the theory of differentiable stable operators, namely, Theorem 44 and Theorem 45.

Appendix A

Topics in Analysis

In this chapter we will summarize a collection of mathematical facts that are the basic for the understanding of our propositions and theorems. We briefly describe some standard concepts and results from analysis and optimization theory.

Recall that an *open mapping* is one that carries open sets to open sets.

Theorem 49 (Banach Open Mapping Theorem) *A bounded linear operator $\mathcal{A}$ from a Banach space onto an other Banach space is an open mapping. Consequently, if it is also one-to-one, then there exists the bounded inverse operator $\mathcal{A}^{-1}$.*

Definition 20 *The space W^* of all continuous linear functionals on a topological vector space W is called the (topological) dual of W. A dual system is a pair $\{W, W^*\}$ of vector spaces together with a function $(x, x^*) \to \langle x, x^* \rangle_{(W^*, W)}$ (called pairing or duality of the pair), satisfying:*

- *the mapping $x^* \to \langle x, x^* \rangle_{(W^*, W)}$ is linear for each $x \in W$;*

- *the mapping $x \to \langle x, x^* \rangle_{(W^*, W)}$ is linear for each $x^* \in W^*$;*

- *if $\langle x, x^* \rangle_{(W^*, W)} = 0$ for each $x^* \in W^*$, then $x = 0_W$;*

- *if $\langle x, x^* \rangle_{(W^*, W)} = 0$ for each $x \in W$, then $x^* = 0_{W^*}$;*

The (norm) dual of a normed space $\{W, \|\cdot\|_W\}$ is the vector space $\{W^*, \|\cdot\|_{W^*}\}$ consisting of all continuous linear functionals on W, equipped with the operator norm $\|\cdot\|_{W^*}$. The norm dual W^{**} of W^* is called the second dual.

Theorem 50 *The norm dual of a normed space is a Banach space.*

Definition 21 *A Banach space W is called reflexive if $W = W^{**}$.*

Note that W is reflexive if and only if W^* is reflexive. Moreover, the reflexivity of W is equivalent to the compactness of the closed unit ball

$$V_1(0) := \{x \in W \; : \; \|x\|_W \leq 1\}$$

in the weak topology.

Theorem 51 (Riesz Theorem) *Let H be a real Hilbert space and H^* be the dual to H. For any functional $l \in H^*$ there exists a unique element $\tilde{h} \in H$ such that*

$$l(h) = \langle \tilde{h}, h \rangle_H \; \forall h \in H.$$

Let H be a Hilbert space. Consider a closed subspace K of H. The set

$$\{h \in H \; : \; \langle h, h_K \rangle_H = 0 \; \forall h_K \in K\}$$

is called an orthogonal complement of K and is denoted by $K^{\perp}$.

Theorem 52 *Let W be a real normed space. Consider a functional $J : W \to \mathbb{R}$. If $J(\cdot)$ is once continuous Fréchet differentiable, then for any $x, y \in W$,*

$$J(x) - J(y) = J_x(x + \lambda(y - x))(y - x)$$

for some $\lambda \in [0, 1]$, where $J_x(\cdot)$ is the Fréchet derivative. If W is a real Hilbert space, then

$$J(x) - J(y) = \langle J_x(x + \lambda(y - x)), (y - x) \rangle_H$$

for some $\lambda \in [0, 1]$.

Theorem 53 (Classic Implicit Function Theorem) *Let X, Y, Z be real Banach spaces and V be a neighborhood in $X \times Y$ and $\Psi : V \to Z$ be a continuously Fréchet differentiable mapping. Assume that*

- $\Psi(x_1, y_1) = 0$;

- *there exists an inverse operator*

$$[\Psi'_y(x_1, y_1)]^{-1} \in L(Z, Y),$$

where Ψ'_y is the Fréchet derivative of $\Psi(x, \cdot)$.

Then there exist $\epsilon, \delta > 0$ and there exists a continuously Fréchet differentiable mapping $\psi : B(x_1, \delta) \to Y$, where

$$B(x_1, \delta) := \{x \in X \; : \; \|x - x_1\| < \delta\},$$

such that

- $\psi(x_1) = y_1$;

- *from $\|x - x_1\|_X < \delta$ follows $\|\psi(x) - y_1\|_Y < \epsilon$ and $\Psi(x, \psi(x)) = 0$;*

- *if $(x, y) \in B(x_1, \delta) \times B(y_1, \epsilon)$, where*

$$B(y_1, \epsilon) := \{x \in X \; : \; \|x - x_1\| < \epsilon\},$$

  then $\Psi(x, y) = 0$ implies $y = \psi(x)$;

- *the mapping $\psi(\cdot)$ is Fréchet differentiable and*

$$\psi'_x(x) = -[\Psi'_y(x, \psi(x))]^{-1} \Psi'_x(x, \psi(x)),$$

  where Ψ'_x is the Fréchet derivative of $\Psi(\cdot, y)$ and ψ'_x is the Fréchet derivative of $\psi(\cdot)$.

Definition 22 *A continuous operator $A : W_1 \to W_2$ between two normed spaces is a nonexpansive operator, if*

$$\|A(x_1) - A(x_2)\|_{W_2} \leq \|x_1 - x_2\|_{W_1}$$

for all $x_1, x_2 \in W_1$. This operator is called contractive if there exists a nonnegative number $r > 0$ with the property that

$$\|A(x_1) - A(x_2)\|_{W_2} \leq r\|x_1 - x_2\|_{W_1}$$

for all $x_1, x_2 \in W_1$.

Theorem 54 (Banach Fixed Point Theorem) *Let X be a Banach space and let A be a contraction of X into itself. Then A has a unique fixed point, in the sense that $A(x) = x$ for some $x \in X$.*

Let W be a vector space. A functional $J : W \to \bar{\mathbb{R}} := \mathbb{R} \bigcup \{+\infty\}$ is convex, if for any $x_1, x_2 \in W$ and $\lambda \in [0, 1]$,

$$J(\lambda x_1 + (1 - \lambda)x_2) \leq \lambda J(x_1) + (1 - \lambda)J(x_2).$$

This functional is *strictly convex* if

$$J(\lambda x_1 + (1 - \lambda)x_2) < \lambda J(x_1) + (1 - \lambda)J(x_2).$$

For J the set

$$\mathrm{epi}J := \{(\alpha, x) \in \mathbb{R} \times W \; : \; \alpha \geq J(x)\}$$

is said to be an epigraph of $J(\cdot)$. Usually, for a convex functional

$$J(\cdot) : W \to \bar{\mathbb{R}},$$

$J(\cdot) \neq \infty$, the notation of a *proper convex* functional is used. The set

$$\mathrm{dom}J := \{x \in W \; : \; J(x) < \infty\}$$

is called an *effective domain* of a functional $J(\cdot)$. Recall that a set of all interior points of a convex set S is called the *interior* of S and denoted by $\mathrm{int}\{S\}$.

Definition 23 *A functional $J : W \to \bar{\mathbb{R}}$ is said to be lower semicontinuous if* $\mathrm{epi}J$ *is a closed set.*

Theorem 55 (Mean Value Theorem) *Let H be a real Hilbert space and*

$$J : H \to \bar{\mathbb{R}}$$

is a lower semicontinuous proper functional. Suppose J is Gâteaux differentiable on an open neighborhood that contains the line segment

$$[h_1, h_2] := \{th_1 + (1 - t)h_2 \; : \; 0 \leq t \leq 1\},$$

where $h_1, h_1 \in H$. Then there exists a $h_3 \in [h_1, h_2]$ such that

$$J(h_1) - J(h_2) = \langle J'_G(h_3), h_1 - h_2 \rangle_H,$$

where $J'_G(h_3)$ is the Gâteaux derivative of J at the point h_3.

Consider a functional $J : X \to \bar{\mathbb{R}}$, where X is a real Banach space. Let $U \subseteq X$. A sequence $\{x_i\}$, $i \in \mathbb{N}$ in X is a *minimizing sequence* if

$$\lim_{i \to \infty} J(x_i) = \inf_{x \in U} J(x).$$

The following references may be useful for a detailed study of the basic facts: in theory of *convergence spaces* (or *Fréchet spaces*) [191, 5], and in theory of *Sobolev spaces* [2, 143, 12].

Let X be a real reflexive Banach space and $Q \subseteq X$ be a nonempty closed convex set. Let $J : X \to (-\infty, \infty]$ be a proper convex and lower semicontinuous functional. Consider the convex optimization problem

$$\begin{aligned} &\text{minimize } J(x) \\ &\text{subject to } x \in Q. \end{aligned} \tag{A.1}$$

For (A.1) we formulate the following existence result [92].

Theorem 56 *Let the set Q be bounded in addition to the above assumptions. Then there exists a solution x^{opt} of (A.1). If J is strictly convex, then (A.1) has a unique solution.*

Note that the boundedness condition in Theorem 56 can be replaced by the condition J to be coercive, i.e.,

$$\lim_{\|x\|_X \to \infty} J(x) = \infty$$

for $x \in Q$. The solutions set of (A.1) is a closed convex set [92].

Theorem 57 *Let in addition to the above assumptions the objective functional J be Gâteaux differentiable. The element $x^{opt} \in Q$ is a solution of (A.1) if and only if*

$$\langle \nabla_G J(x^{opt}), x - x^{opt} \rangle_{(X^*,X)} \geq 0 \ \forall x \in X,$$

where $\nabla_G J(x^{opt})$ denotes the Gâteaux derivative of J at x^{opt}.

The central result of the optimization theory describes the first order necessary optimality conditions for the smooth finite-dimensional version of the general nonlinear problem

$$\text{minimize } J(x)$$
$$\text{subject to } x \in S \tag{A.2}$$
$$g_i(x) \leq 0, \ i = 1, ..., m \ , \ h(x) = 0,$$

where $S \subseteq \mathbb{R}^n$ is an open set, $J : \mathbb{R}^n \to \mathbb{R}$ is Fréchet differentiable, $g_i : \mathbb{R}^n \to \mathbb{R}$ for all $i = 1, ..., m$ and $h : \mathbb{R}^n \to \mathbb{R}^r$. Let $x^{opt} \in \mathbb{R}^n$ be a local minimizer of (A.2). We denote the set of indices of the *active* inequality constraints by

$$I(x^{opt}) := \{i \ : \ g_i(x^{opt}) = 0\}.$$

Let the given functions g_i, $i = 1, ..., m$ be continuous and Fréchet differentiable at the point x^{opt}. The function h is assumed to be strictly differentiable, with the surjective gradient, at x^{opt}. Suppose that there is a direction $p \in \mathbb{R}^n$ satisfying

$$\langle \nabla g_i(x^{opt}), p \rangle < 0 \ \forall i \in I(x^{opt}) \tag{A.3}$$

The introduced conditions (A.3) are the Mangasarian-Fromovitz regularity conditions (constraint qualifications).

Theorem 58 (Karush-Kuhn-Tucker) *If the above assumptions and the above Mangasarian-Fromovitz constraint qualifications hold, then there exist the Lagrange multipliers $\lambda_i \in \mathbb{R}_+$ (for $i \in I(x^{opt})$) and $\mu \in \mathbb{R}^n$ satisfying*

$$\nabla J(x^{opt}) + \sum_{i \in I(x^{opt})} \lambda_i \nabla g_i(x^{opt}) + \langle \nabla h(x^{opt}), \mu \rangle = 0.$$

Next we present the known Ekeland Variational Principle in Banach spaces. Note that the corresponding result can also be proved for complete metric spaces [203].

Theorem 59 (Ekeland Variational Principle) *Let X be a real Banach space and let*

$$J : X \to (-\infty, \infty]$$

be a proper, bounded below and lower semicontinuous functional. Then, for any $\epsilon > 0$ and $\tilde{x} \in X$ with

$$J(\tilde{x}) \le \inf_{x \in X} J(x) + \epsilon,$$

there exists $z \in X$ satisfying the following conditions

$$J(z) \le J(\tilde{x}), \quad \|\tilde{x} - z\|_X \le 1,$$
$$J(\xi) > J(z) - \epsilon\|z - \xi\|_X \; \forall \xi \in X, \; \xi \ne z.$$

Let X be a real Banach space and $J : X \to (-\infty, \infty]$ be a functional on X. An element $u^* \in X^*$ is called a *subgradient* of J at $u \in X$ if $J(u) \ne \pm\infty$ and

$$J(x) \ge J(u) + \langle u^*, x - u \rangle_{(X^*, X)} \; \forall x \in X.$$

The set of all subgradients of J at $u \in X$ is called a subdifferential $\partial_S J(x)$ of J at $x \in X$. For a proper convex functional J with $J(x) < \infty$ we have $\partial_S J(x) \ne$ for all $x \in X$ and the set $\partial_S J(x)$ is convex. We refer to [80, 81, 84, 186] for analytical details and for other concepts from the nonsmooth analysis and nonlinear analysis. Using the generalized Jacobian of a mapping $G : D \subseteq \mathbb{R}^n \to \mathbb{R}$, we formulate the next well known theorem.

Theorem 60 (Inverse Function Theorem) *Let $D \subseteq \mathbb{R}^n$ be an open set,*

$$G : D \to \mathbb{R}^n$$

be a locally Lipschitz mapping, $x_0 \in D$. Assume that all linear operators from the generalized Jacobian $\partial G(x_0)$ are invertible. Put $G(x_0) = y_0$. Then there exist $\eta > 0 \; \delta > 0$ and a Lipschizt mapping G^{-1} defined on an open ball $\mathrm{int}\{V_\eta(y_0)\}$ such that

$$G(G^{-1}(y)) = y \; \forall y \in \mathrm{int}\{V_\eta(y_0)\},$$
$$G^{-1}(G(x)) = x \; \forall x \in \mathrm{int}\{V_\delta(x_0)\}.$$

Let us now consider a fundamental principle of the linear functional analysis and two consequences.

Theorem 61 (Hahn-Banach Theorem) *Let M be a subspace of a given normed space W, and let Λ_0 be a continuous linear functional on M. Then Λ_0 can be extended to a continuous linear functional Λ_1 defined on the whole space W such that*

$$\|\Lambda_0\|_W = \|\Lambda_1\|_W.$$

Theorem 62 (Separation Theorem) *Let W be a normed space and let C be a nonempty closed convex subset of W. If x is a vector not in C, then there exists a continuous linear functional $\Lambda \in W^*$ such that*

$$\Lambda(x) < \inf_{y \in W} \Lambda(y).$$

Theorem 63 *Let X be a real Banach space and let $J : X \to (-\infty, \infty]$ be a proper convex and lower semicontinuous functional. Then there are $e \in \mathbb{R}$ and $x^* \in X^*$ such that for all $x \in X$*

$$J(x) \geq e + \langle x^*, x \rangle_{(X^*, X)}$$

If $S \subseteq X$ is a closed convex set such that $J : S \to \mathbb{R}$, then J is continuous on the interior int$\{S\}$.

The Liusternik Theorem is a basic of some proofs presented in this book. In connection with this we formulate two classical results and refer to [123, 60] for some generalizations of the Liusternik's result.

Theorem 64 (Liusternik Theorem) *Let X, Y be real Banach spaces and let*

$$A : X \to Y$$

be a strictly Fréchet differentiable operator at $x_0 \in X$. Assume that

$$M := \{x \in X \ : \ A(x) = 0\} \neq \emptyset.$$

Then there exists a tangent space to the set M at the point $x_0 \in M$. If the operator $A'(x_0)$ is surjective, then

$$T_z(M) = T_z^+(M) = \text{Ker}(A'(z))$$

for all $z \in M$, where Ker$(A'(z)) := \{x \in X \ : \ A'(z)x = 0\}$ *is the kernel of $A'(z)$.*

Theorem 65 (Graves-Liusternik Theorem) *Let H_1, H_2 be two real Hilbert spaces, W be a metric space, $U \subseteq H_1$ and $F : H_1 \times W \to H_2$ be a partial Fréchet differentiable mapping. Let $(h_0, \omega_0) \in U \times W$ be a point satisfying $F(h_0, \omega_0) = 0$, and suppose that $F'_h(h_0, \omega_0)$ is onto:*

$$R(F'_h(h_0, \omega_0)) = H_2.$$

Let Ω be a neighborhood of h_0. Then for some $\delta > 0$, for all (h, ω) sufficiently near (h_0, ω_0), we have

$$\text{dist}(\Phi(\omega), h) \leq \frac{\|F(h, \omega)\|_{H_2}}{\delta},$$

where $\Phi(\omega) := \{z \in \Omega \ : \ F(z, \omega) = 0\}$.

Let X be a real Banach space and $S \subset X$ be closed and convex. We consider the following *variational inequality*

$$\langle f(\omega, \tilde{x}), x - \tilde{x} \rangle_{(X^*, X)} \geq 0 \ \forall x \in S, \tag{A.4}$$

where $f : W \times X \to X^*$ and W is a normed space (a parameter set). Let us introduce the *normal cone* to S at $x \in X$

$$N_S(x) := \begin{cases} \{v \in X \ : \ \langle v, x - \tilde{x} \rangle_{(X^*, X)} \leq 0 \ \forall x \in S\}, & \text{if } \tilde{x} \in S; \\ \emptyset, & \text{otherwise.} \end{cases}$$

We assume that the given variational inequality (A.4) has a solution $x_1 \in X$ for a fixed parameter $\omega_1 \in W$. Let $f(\cdot, x_1)$ and $f'_x(\cdot, x_1)$ be continuous at ω_1 and $f(\omega, \cdot)$ be Fréchet differentiable. Let $f'_x(\omega, \cdot)$ be continuous uniformly in $\omega \in W$. The next theorem is of importance in explaining the regularity conditions in the general optimization theory.

Theorem 66 (Robinson Implicit Function Theorem) *Let the conditions given above hold. Assume that there exist a neighborhood Ξ of x_1 and Θ of the origin such that the mapping*

$$x \to [f(\omega_1, x_1) + f'_x(\omega_1, x_1)(\cdot - x_1) + N_S(\cdot)](x) \bigcap \Xi$$

is single-valued and Lipschitz continuous in Θ with a Lipschitz constant L. Then for every $\mu > 0$ there exist neighborhoods $\mathcal{V}_\mu$ of x_1 and $\mathcal{U}_\mu$ and a single-valued function $\omega \to x(\omega)$ from $\mathcal{U}_\mu$ to $\mathcal{V}_\mu$ such that for any $\omega \in \mathcal{U}_\mu$, $x(\mu)$ is a unique solution in $\mathcal{V}_\mu$ of (A.4) and moreover, for any $\omega, \tilde{\omega} \in \mathcal{V}_\mu$ one has

$$\|x(\omega) - x(\tilde{\omega})\|_X \leq (L + \mu)\|f(\omega, x(\tilde{\omega})) - f(\tilde{\omega}, x(\tilde{\omega}))\|_{X^*}.$$

We refer to [86] for some interesting generalization of Theorem 65. Consider a proper convex and lower semicontinuous functional

$$J : X \to (-\infty, \infty]$$

on the real Banach space X. The *conjugate functional* is given by

$$J^*(p) := \sup_{x \in X} \langle p, x \rangle_{X^*, X}.$$

The next result establishes the relationship between Tikhonov well-posedness and differentiability of the conjugate functional [11].

Theorem 67 (Asplund-Rockafellar Theorem) *Let X be a real Banach space. The problem*

$$\text{minimize } J(x)$$
$$\text{subject to } x \in X$$

is Tikhonov well-posed with respect to the strong convergence if and only if J^* is Fréchet differentiable at 0.

Let W be a vector space and $Q \subset W$ be a convex set. A function

$$f : Q \to (-\infty, \infty]$$

is convex on W if for any $w_1, w_2 \in W$ and $\alpha \in [0, 1]$,

$$f(\alpha w_1 + (1 - \alpha)w_2) \leq \alpha f(w_1) + (1 - \alpha)f(w_2).$$

We now consider some properties of convex functions.

Theorem 68 *Let Q be a convex subset of a vector space W and let*

$$f : Q \to (-\infty, \infty]$$

be a convex function on W. Then, for any $c \in \mathbb{R}$,

$$G_c := \{w \in W \ : \ f(x) \leq c\}$$

is convex.

Theorem 69 *Let Q be a convex subset of a vector space W, let f, g be convex functions on W and let c ∈ ℝ₊. Then two functions f + g and cf defined by*

$$(f + g)(w) = f(w) + g(w), \quad (cf)(w) = cf(w)$$

for all w ∈ W are convex on W.

Theorem 70 *Let Q be a closed convex subset of a Banach space X and let f be a convex function of Q into (−∞, ∞]. Then, f is lower semicontinuous in the norm topology if and only if f is lower semicontinuous in the weak topology.*

We formulate here the existence result for the convex optimization problem given above as a theorem.

Theorem 71 *Let Q be a closed convex subset of a reflexive Banach space X. Let f be a proper convex lower semicontinuous function of Q into (−∞, ∞] and suppose that*

$$\lim_{n \to \infty} f(x_n) = \infty$$

as $\lim_{n \to \infty} \|x_n\|_X = \infty$. *Then there exists an element* $x_0 \in D(f)$ *such that*

$$f(x_0) = \inf_{x \in Q} f(x).$$

Finally, we present some useful property of strongly convex functions. Let D be a nonempty convex set in $\mathbb{R}^n$. Recall that a function

$$\Gamma : D \subseteq \mathbb{R}^n \to \mathbb{R}$$

is called strongly convex on D with modulus $\gamma > 0$ if

$$\Gamma(\alpha x_1 + (1 - \alpha)x_2) \le \alpha\Gamma(x_1) + (1 - \alpha)\Gamma(x_2) - \frac{\gamma}{2}\alpha(1 - \alpha)\|x_1 - x_2\|^2$$

for all $x_1, x_2 \in D$ and all $\alpha \in]0, 1[$. Note that the function Γ is strongly convex on D with modulus γ if and only if the function

$$\Gamma(\cdot) - \frac{\gamma}{2}\| \cdot \|^2$$

is convex on D.

Theorem 72 *Let Γ be a function differentiable on an open set*

$$\Omega \subset \mathbb{R}^n,$$

and let D be a convex subset of Ω. Then Γ is strongly convex with modulus γ on D if and only if, for all $x_1, x_2 \in D$,

$$\Gamma(x_1) - \Gamma(x_2) \geq \langle \nabla\Gamma(x_1), x_1 - x_2 \rangle + \frac{\gamma}{2}\|x_1 - x_2\|^2.$$

The gradient of a convex function is a monotone operator and the gradient of the strongly convex function Γ is strongly monotone (strongly stable), i.e.,

$$\langle \nabla\Gamma(x_1) - \nabla\Gamma(x_2), x_1 - x_2 \rangle \geq \gamma\|x_1 - x_2\|^2.$$

Evidently, $\nabla\Gamma(\cdot)$ is an expanding operator.

Bibliography

[1] Abraham, R., *Foundations of Mechanics*,WA Benjamin, New York, 1967.

[2] Adams, R.A., *Sobolev Spaces*, Academic Press, New York, 1973.

[3] Alber, Y.I., Burachik, R.S. and Iusem, A.N., A proximal point method for nonsmooth convex optimization problems in Banach spaces, *Abstract and Applied Analysis* **2** (1997), 97-120.

[4] Alekseev, V.M., Tichomirov, V.M. and Fomin, S.V., *Optimal Control*, Plenum Publishing Co., New York, 1987.

[5] Aliprantis, C.D. and Border, K.C., *Infinite Dimensional Analysis*, Springer, New York, 1999.

[6] Alvarez, F., On the minimizing property of a second order dissipative system in Hilbert spaces, *SIAM Journal on Control and Optimization* **38** (2000), 1102-1119.

[7] Amato, F., *Robust Control of Linear Systems Subject to Time-Varying Parameters*, Springer, Berlin, 2005.

[8] Angeli, D. and Sontag, E.D., Monotone control systems, *IEEE Transactions on Automatic Control* **48** (2003), 1684-1698.

[9] Arnold, V.I., *Mathematical Methods of Classical Mechanics* (translation of the 1974 Russian edition), Springer, Berlin, 1978.

[10] Arutyunov, A.V. and Aseev, S.M., Investigation of the degeneracy phenomenon in the maximum principle for optimal control with state constraints, *SIAM Journal on Control and Optimization* **35** (1997), 930-952.

[11] Asplund, E. and Rockafellar, R., Gradients of convex functions, *Trans. Amer. Math. Soc.* **139** (1969), 443-467.

[12] Atkinson, K. and Han, W., *Theoretical Numerical Analysis*, Springer, New York, 2005.

[13] Auslender, A., Numerical methods for nondifferentiable convex optimization, *Mathematical Programming Study* **30** (1987), 102-126.

[14] Auslender, A., Crouzeix, J.P. and Fedit, P., Penalty proximal methods in convex programming, *Journal of Optimization Thery and Applications* **55** (1987), 1-21.

[15] Auslender, A., Regularity theorems in sensitivity theory with nonsmooth data. In: *Parametric Optimization and Related Topics* (J. Guddat, H.Th. Jongen, B. Kummer and F. Nozicka, eds.), Akademie-Verlag, Berlin, pp.9-15, 1987.

[16] Azhmyakov, V., A constructive method for solving stabilization problems, *Discussiones Mathematicae Differential Inclusions, Control and Optimization* **20** (2000), 51-62.

[17] Azhmyakov, V., Optimal control of well-stired bioreactor in the presence of stochastic perturbations, *Informatica* **13** (2002), 133-148.

[18] Azhmyakov, V. and Schmidt, W.H., Strong convergence of a proximal-based method for convex optimization, *Mathematical Methods of Operations Research* **57** (2003), 393-407.

[19] Azhmyakov, V. and Schmidt, W.H., Explicit approximations of relaxed optimal control processes. In: *Optimal Control* (W.H. Schmidt and G. Sachs, eds.), Hieronymus Bücherproduktion GmbH, München, pp.179-192, 2003.

[20] Azhmyakov, V., A numerically stable method for convex optimal control problems, *Journal of Nonlinear and Convex Analysis* **5** (2004), 1-18.

[21] Azhmyakov, V. and Schmidt, W.H., On the optimal design of elastic beams, *Structural and Multidisciplinary Optimization*, **27** (2004), 80-88.

[22] Azhmyakov, V., A numerical method for optimal control problems using proximal point approach, *Aportaciones Matematicas* **18** (2004), 13-31.

[23] Azhmyakov, V., A proximal-based method for relaxed optimal control problems with constraints. In: *Mesh Methods for Boundary-Value Problems and Applications* (I.B. Badriev and F.M. Ablaev, eds.), Kazan State University Press, Kazan, pp.249-254, 2004.

[24] Azhmyakov, V and Schmidt, W.H., Relaxed optimal control problems with constraints: a computational approach, *Preprint-Reihe Mathematik University of Greifswald*, Greifswald, 2004.

[25] Azhmyakov, V., *Stable Operators in Analysis and Optimization*, Peter Lang, Berlin, 2005.

[26] Azhmyakov, V., Stable methods for convex optimal control problems with constraints, *Preprint-Reihe Mathematik University of Greifswald*, Greifswald, 2005.

[27] Azhmyakov, V. and Schmidt, W.H, Approximations of relaxed optimal control problems, *Journal of Optimization Theory and Applications* **130** (2006), 61-78.

[28] Azhmyakov, V., Stability of differential inclusions: a computational approach, *Mathematical Problems in Engineering* **1** (2006), 1-15.

[29] Azhmyakov, V., On differentiable stable operators in Banach spaces, *Fixed point Theory and Applications* **2** (2006), 1-17.

[30] Azhmyakov, V. and Raisch, J. Convex control systems and convex optimal control problems with constraints, *IEEE Transactions on Automatic Control*, to appear.

[31] Azhmyakov, V. and Raisch, J., A gradient-based approach to a class of hybrid optimal control problems. In: *Proceedings of the 2nd IFAC Conf. on ADHS*, Alghero, pp. 89-94, 2006.

[32] Azhmyakov, V., Attia, S.A., Gromov, D. and Raisch, J., Necessary optimality conditions for a class of hybrid optimal control problems. In: LNCS 4416, pp.637-640, 2007.

[33] Attia, S.A., Azhmyakov, V. and Raisch, J., State jump optimization for a class of hybrid autonomous systems. In: *Proceedings of the CCA 2007*, to appear.

[34] Azhmyakov, V., A computational approach to optimization of controlled mechanical systems. In: *proceedings of the MED 2007*, to appear.

[35] Azhmyakov, V., Optimal control in mechanics. *Differential Equations and Nonlinear Mechanics*, accepted.

[36] Baillieul, J., The geometry of controlled mechanical systems. In: *Mathematical Control Theory* J. Baillieul and J.C. Willems, Eds. New York, NY: Springer, pp. 322-354, 1999.

[37] Bakushinskii, A.B., Solution methods for monotonous variational inequalities founded on the principle of iterative regularization, *U.S.S.R. Computational Mathematics and Mathematical Physics* **17** (1977), 12-24.

[38] Balakrishnan, A.V., On a new computing technique in optimal control and its application to minimal-time flight profile optimization, *Journal of Optimization Theory and Applications* **4** (1969),

[39] Balakrishnan, A.V. and Neustadt, L.W. (eds.), *Conference on Computational Methods in Optimization Problems*, Academic Press, New York, 1964.

[40] Bauschke, H.H., Borwein, J.M. and Combettes, P.L., Bregman monotone optimization algorithms, Preprint, 2002.

[41] Bauschke, H.H., Borwein, J.M. and Combettes, P.L., Bregman monotone optimization algorithms, *SIAM Journal on Control and Optimization* **42** (2003), 596-636.

[42] Bellman, R. and Dreyfus, S.E., *Applied Dynamic Programming*, Princeton University Press, Princenton, 1962.

[43] Benker, H., Hamel, A. and Tammer, C., A proximal point algorithm for control approximation problems, *Mathematical Methods of Operations Research* **43** (1996), 261-280.

[44] Berkovitz, L.D., *Optimal Control Theory*, Springer, New York, 1974.

[45] Bertsekas, D.P., *Constrained Optimization and Lagrange Multiplier Method*, Academic Press, New York, 1982.

[46] Bertsekas, D.P., *Dynamic Programming and Optimal Control*, Athena Scientific, Belmont, 1995.

[47] Betts, J.T., Using sparse nonlinear programming to compute low trust orbit transfers, *Boing Computer Services Technical Report*, 1992.

[48] Betts, J.T., *Practical Methods for Optimal Control Using Nonlinear Programming*, SIAM, Philadelphia, 2001.

[49] Bittner, L., New conditions for validity of the Lagrange multiplier rule, *Mathematische Nachrichten* **48** (1971), 353-370.

[50] Bittner, L., (1976). Necessary Optimality Conditions for a Model of Optimal Control Processes. In: *Mathematical Control Theory* (C. Olech and E. Fidelis, eds.), Polish Scientific Publications, pp.25-32, 1996.

[51] A.M. Bloch and P.E. Crouch, Optimal control, optimization and analytical mechanics. In: *Mathematical Control Theory* J. Baillieul and J.C. Willems, Eds. New York, NY: Springer, pp. 268-321, 1999.

[52] Bogatyreva, N.A. and Pyatnitskii, Y.S., The minimax principle in mechanics, *Journal of Applied Mathematics and Mechanics* **57** (1993), 761-770.

[53] Bogoljubov, N.N., Sur quelnes methods nouvelles dans le calculus des variations, *Ann. Math. Pura Appl.* **7** (1930), 249-271.

[54] Boltyanski, V., Martini, H. and Soltan, V., *Geometric Methods and optimization Problems*, Kluver, Dordrecht, 1999.

[55] Bonnans, J.F., On an algorithm for optimal control using Pontryagin's maximum principle, *SIAM Journal of Control and Optimization* **24** (1986), 579-588.

[56] Borwein, J.M. and Lewis, A.S., *Convex Analysis and Nonlinear Optimization*, Springer, New York, 2000.

[57] Branicky, M.S., V.S. Borkar and S.K. Mitter (1998). A unifed framework for hybrid control: model and optimal control theory. *IEEE Trans. Automat. Contr.* **43**, pp. 31-45.

[58] Breakwell, J.V., The optimization of trajectories, *SIAM Journal* **7** (1959), 215-247.

[59] A. Bressan, "Impulsive control systems", in *Nonsmooth Analysis and Geometric Methods in Deterministic Optimal Control* B. Mordukhovich and H.J. Sussmann, Eds. New York, NY: Springer, pp. 1-22, 1996.

[60] Brezhneva, O. and Tret'yakov, A.A., Optimality conditions for degenerate extremum problems with equality constraints, *SIAM J. Control Optim.* **2** (2003), 729-745.

[61] R.W. Brockett, "Control theory and analytical mechanics", in *Geometric Control Theory* C.F. Martin and R. Hermann, Eds. Brooklyn, NY: Math. Sci. Press, 1976.

[62] Browder, F.E., Convergence of approximants to fixed points of non-expansive non-linear mappings in Banach spaces, *Archive for Rational Mechanics and Analysis* **24** (1967), 82-90.

[63] Bryson, A.E. and Ho, Y.C., *Applied Optimal Control*, Wiley, New York, 1975.

[64] Bryson, A.E. and Denham, W.F., A steepest ascent method for silving optimum programming problems, *Journal of Appl. Mechanics* **29** (1962), 247-257.

[65] Bulirsch, R., Montrone, F. and Pesch, H.J., Abort landing in the presence of windshear as a minimax optimal control problem, Part 1: Necessary conditions, *Journal of Optimization Theory and Applications* **70** (1991), 1-23.

[66] Bulirsch, R., Montrone, F. and Pesch, H.J., Abort landing in the presence of windshear as a minimax optimal control problem, Part 2: Multiple chooting and homotopy, *Journal of Optimization Theory and Applications* **70** (1991), 223-254.

[67] Burachik, R.S. and Iusem, A.N., A generalized proximal point algorithm for the variational inequality problem in a Hilbert space, *SIAM Journal on Optimization* **8** (1998), 197-216.

[68] Burachik, R.S., Butnariu, D. and Iusem, A.N., Iterative methods for solving stochastic convex feasibility problems and applications, *Computational Optimization and Applications* **15** (2000), 269-307.

[69] Burachik, R.S. and Scheimberg, S., A proximal point method for the variational inequality problem in Banach spaces, *SIAM Journal on Control and Optimization* **39** (2001), 1633-1649.

[70] Butnariu, D., Iusem, A.N., On a proximal point method for convex optimization in Banach spaces, *Numerical Functional Analysis and Optimization* **18** (1997), 723-744.

[71] Büskens, C. and Maurer, H., SQP-methods for solving optimal control problems with control and state constraints: adjoint variables, sensitivity analysis and real-time control, *Journal of Computational and Applied Mathematics* **120** (2000), 85-108.

[72] Butzek, S. and Schmidt, W.H., Relaxation gaps in optimal control processes with state constraints. In: *Variational Calculus Optimal Control and Applications*, (W.H. Schmidt, K. Heier, L. Bittner and R. Bulirsch, eds.) Birkhäuser, Basel, Basel, pp. 21-29, 1998.

[73] Caines, P. and Shaikh, M.S., Optimality zone algorithms for hybrid systems computation and control: From exponential to linear complexity. In: *Proceedings of the 13th Mediterranean Conference on Control and Automation.* pp. 1292-1297, 2005.

[74] Cassandras, C., Pepyne, D.L. and Wardi, Y., Optimal control of a class of hybrid systems. *IEEE Trans. Automat. Contr.* **46** (2001), 398-415.

[75] Censor, Y. and Zenios, S.A., The proximal minimization algorithm with D-functions, *Journal of Optimization Theory and Applications* **73** (1992), 451-464.

[76] Cesari, L., *Optimization Theory and Applications*, Springer, New York, 1983.

[77] Chernousko, F.L. and Lyubuschin, A.A., Method of successive approximations for solution of optimal control problems, *Optimal Control, Appl. and Methods* **3** (1982), 101-114.

[78] Clarke, F.H., The generalized problem of Bolza, *SIAM Journal on Control and Optimization* **14** (1976), 683-699.

[79] Clarke, F.H., Optimal solutions to differential inclusions, *Journal of Optimization Theory and Applications* **19** (1976), 469-478.

[80] Clarke, F.H., *Optimization and Nonsmooth Analysis*, SIAM, Philadelphia, 1990.

[81] Clarke, F.H., Yu.S. Ledyaev, R.J.Stern and P.R. Wolenski, *Nonsmooth Analysis and Control Theory*, Springer, New York, 1998.

[82] Cullum, J., An explicit procedure for discretizing continuous optimal control problems, *Journal of Optimization Theory and Applications* **8** (1971), 15-34.

[83] Daniel, J.W., On the convergence of a numerical method in optimal control, *Journal of Optimization Theory and Applications* **4** (1969), 330-342.

[84] V.F. Demyanov and A.M. Rubinov, *Constructive Nonsmooth Analysis*, Peter Lang, Frankfurt, 1995.

[85] Dontchev, A.L. and Lempio, F., Difference methods for differential inclusions: a survey, *SIAM Review* **34** (1992), 263-294.

[86] Dokov, S.P. and Dontchev, A., Robinson's strong regularity implies robust local convergence of Newton's method. In: (W.H. Hager and P.M. Pardalos), *Optimal Control*, Kluver, Dordrecht, 116-129, 1998.

[87] Dontchev, A. and Zolezzi, T., *Well Posed Optimization Problems*, Springer, Berlin, 1993.

[88] Dontchev, A.L. and Hager, W.W., Lipschitz stability in nonlinear control and optimization, *SIAM Journal on Control and Optimization* **31** (1993), pp.569-603.

[89] Dontchev, A.L., Discrete Approximations in Optimal Control. In: *Nonsmooth Analysis and Geometric Methods in Deterministic Optimal Control* (B.S. Mordukhovich and H.J. Sussmann, eds.), Springer, New-York, pp. 59-80, 1996.

[90] Dunn, J.C., On state constraint representations and mesh-depent gradient projection convergence rates for optimal control problems, *SIAM Journal of Control and Optimization* **39** (2000), 1082-1111.

[91] Dunn, J.C., Diagonally modified conditional gradient method for input constrained optimal control problems, *SIAM Journal of Control and Optimization* **24** (1986), 1177-1191.

[92] Ekeland, I. and Temam, R., *Convex Analysis and Variational Problems*, North-Holland, Amsterdam, 1976.

[93] Esposito, W.R. and Floudas, C.A., Deterministic global optimization in nonlinear optimal control problems, *Journal of Global Optimization* **17** (2000), 97-126.

[94] Fattorini, H.O., *Infinite Dimensional Optimization and Control Theory*, Cambridge University Press, Cambridge, 1999.

[95] Fedorenko, R.P., *Priblizhyonnoye Reshenyie Zadach Optimalnogo Upravlenya* (in Russian), Nauka, Moscow, 1978.

[96] Ferreira, M.M.A., Fontes, F.A.C.C. and Vinter, R.B., Nondegenerate necessary conditions for nonconvex optimal control problems with state constraints, *Journal of Mathematical Analysis and Applications* **233** (1999), 116-129.

[97] Fiacco, A.V. and McCormick, G, *Nonlinear Programming: Sequential Unconstrained Minimization Techniques*, Wiley, New York, 1968.

[98] Filippov, A.F., On Certain questions in the theory of optimal control (in Russian), *Vestnik Moskovskovo Universiteta* **2** (1959), 25-32. English translation: *SIAM Journal on Control* **1** (1962), 76-84.

[99] Filippov, A.F. (1988). *Differential Equations with Discontinuous Right-Hand Sides*. Kluwer Academic Publishers, Dordrecht.

[100] Fiacco, A.V. and G. McCormick, (1968). *Nonlinear Programming: Sequential Unconstrained Minimization Techniques*, Wiley, New York.

[101] Fletcher, R., *Practical Optimization*, J. Wiley, Chichester, 1989.

[102] Fukushima, M. and Mline, H., A generalized proximal point algrithm for certain non-convex minimization prolems, *International Journal of Systems Science* **12** (1981), 989-1000.

[103] Fukushima, M. and Yamamoto, Y., A second-order algorithm for continuous-time nonlinear optimal control problems, *IEEE Transaction on Automatic Control* **31** (1986), 673-676.

[104] Fuller, A.T., Relay control systems optimized for various performance criteria, *Automatic and Remote Control* **1** (1961), 510-519.

[105] Gabasov, R. and Kirillova, F., *Qualitative Theory of Optimal Processes* (in Russian), Nauka, Moscow, 1971.

[106] Gamkrelidze, R., *Principles of Optimal Control Theory*, Plenum Press, London, 1978.

[107] Gantmakher, F.R., *Lectures on Analytical Mechanics* (in Russian). Moscow, Russia: Nauka 1966.

[108] Ginsburg, B. and Ioffe, A., The maximum principle in optimal control of system governed by semilinear equations. In: *Nonsmooth Analysis and Geometric Methods in Deterministic Optimal Control* (B.S. Mordukhovich and H.J. Sussmann, eds.), Springer-Verlag, New-York, pp.81-110, 1996.

[109] Goddard, R.H., A method for reaching extreme altitude, *Smithsonian Collection* **2** (1919).

[110] Goh, C.J. and Teo, K.L., Control parametrization: a unifed approach to optimal control problems with general constraints, *Automatica* **24** (1988), 3-18.

[111] Goldstein, A.A., Convex programming in Hilbert space, *Bull. Amer. Math. Soc.* **70** (1964), 709-710.

[112] Göpfert, A., Tammer, C. and Riahi, H., Existence and proximal point algorithms for nonlinear monotone complementarity problems, *Optimization* **45** (1999), 57-68.

[113] Güler, O., On the convergence of the proximal point algorithm for convex minimization, *SIAM Journal on Control and Optimization* **29** (1991), 403-419.

[114] Hadamard, J., *Lectures on Cauchy Problems in Linear Partial Differential Equations*, Yale University Press, New Haven, 1923.

[115] Hager, W.W., Rate of convergence for discrete approximations to unconstrained control problems, *SIAM Journal on Numerical Analysis* **13** (1976), 449-471.

[116] Hager, W.W. and Ianculescu, G.D., Dual approximations in optimal control, *SIAM Journal on Control and Optimization* **22** (1990), 1061-1080.

[117] Hestenes, M., *Conjugate Direction Methods in Optimization*, Springer, New York, 1980.

[118] Hiriart-Urruty, J.B. and Lemarechal, C., *Convex Analysis and Minimization Algorithms*, Springer, Berlin, 1993.

[119] Hofmann, B., Ill-posedness and local ill-posedness concepts in Hilbert spaces, *Optimization* **48** (2000), 219-238.

[120] Ioffe, A.D. and Tichomirov, V.M., *Theory of Extremal Problems*, North Holland, Amsterdam, 1979.

[121] Iusem, A.N., Inexact version of proximal point and augmented Lagrangian algorithms in Banach spaces, *Numerical Functional Analysis and Optimization* **22** (2001), 609-640.

[122] Izmailov, A.F. and Solodov, M.V., Optimality conditions for irregular inequality-constrained problems, *SIAM Journal on Control and Optimization* **40** (2001), 1280-1295.

[123] Izmailov, A.F. and Tretjakov, A.A., *2-Regular Solutions of Nonlinear Problems*, Nauka, Moscow, 1999.

[124] Jahn, J., *An Introduction to the Theory of Nonlinear Optimization*, Springer, Berlin, 1974.

[125] Kantorovich, L.V. and Akilov, G.P., *Functional Analysis*, Pergamon Press, Oxford, 1982.

[126] Kantorovich, L.V. and Akilov, G.P., *Functional Analysis* (in Russian), Nauka, Moscow, 1982.

[127] Kaplan, A. and Tichatschke, R., *Stable Methds for Ill-Posed Variational Prolems - Prox-Regularization of Elliptical Variational Inequalities and Semi-Infinite Optimization Problems*, Akademie Verlag, Berlin, 1994.

[128] Kaplan, A. and Tichatschke, R., Proximal point approach and approximation of variational inequalities, *SIAM Journal on Control and Optimization* **39** (2000), 1136-1159.

[129] Kaplan, A. and Tichatschke, R., A general view on proximal point methods to variational inequalities in Hilbert spaces-iterative regularization and approximation, *Journal of Nonlinear and Convex Analysis* **2** (2001), 305-332.

[130] Kaplan, A. and Tichatschke, R., Proximal point methods and nonconvex optimization, *Journal of Global Optimization* **13** (1998), 389-406.

[131] Kelley, C.T. and Sachs, E.W., Mesh independence of the gradient projection method for optimal control problems, *SIAM Journal of Control and Optimization* **30** (1992), 477-493.

[132] Köthe, G., *Topological Vector Spaces*, Springer, Berlin, 1983.

[133] Krasnoselskii, M.A., Solutions of equations involving adjoint operatorsby successive approximations, *Uspekhi Mat. Nauk* **15** (1960), 161-165.

[134] Kress, R., *Linear Integral Equations*, Springer, Berlin, 1989.

[135] Kryanev, A.V., The solution of incorectly posed problems by methods of successive approximations, *Soviet Math. Dokl.* **14** (1973), 673-676.

[136] Kupfer, F.S. and Sachs, E., Reduced SQP method for nonlinear heat conduction control problems, *International Series of Numerical Mathematics* **111** (1993), 145-160.

[137] Kuratowski, K., *Topology*, Academic Press, New York, 1968.

[138] Ledzewicz, U. and Schättler H., A hight-order generalization of the Lysternik theorem, *Nonlinear Anal.* **34** (1998), 793-815.

[139] Lehdili, N. and Moudafi, A., Combining the proximal algorithm and Tikhonov regularization, Optimization **37** (1996), 239-252.

[140] Leibfritz, F. and Sachs, E., Optimal static output feedback design using a trust region interior point method. Presentation at the First Workshop on Nonlinear Optimization *Interior Point and Filter Methods*, Coimbra, Portugal, 1999.

[141] Lemaire, B., On the convergence of some iterative methods for convex minimization. In: *Recent Developments in Optimization* (P. Gritzmann, R. Horst, E. Sachs and R. Tichatschke, eds.), Springer, Berlin, pp.252-268, 1995.

[142] Lemaire, B., Regularization of fixed-point problems and staircase iteration. In: *Ill-Posed Variational Problems and Regularization Techniques* (M. Thera and R. Tichatschke, eds.), Springer, Berlin, pp.151-166, 1998.

[143] Lions, J.P., *Controle Optimal de Systèmes Gouvernés par des Équations aux Dérivées Partielles*, Dunod Gauthier-Villars, Paris, 1968.

[144] Lucchetti, F., Patrone, A., A characterization of Tikhonov well posedness for minimum problems with application to variational inequalities, *Numerical Functional Analysis and Optimization* **3** (1981), 461-476.

[145] Lucchetti, F., Patrone, A., Some properties of "well-posed" variational inequalities governed by linear operators, *Numerical Functional Analysis and Optimization* **5** (1982-1983), 349-361.

[146] Luenberger, D.G., *Investment science*, Oxford University Press, New York, 1998.

[147] Luus, R. and Bojkov, B., Global optimization of bifunctional catalyst process, *Canadian Journal of Chemical Engineering* **72** (1994), 160-163.

[148] Machielsen, K.C.P., *Numerical Solution of Optimal Control Problems with State Constraints by Sequential Quadratic Programming in Function Space*, Thesis, Tecnical University of Eindhoven, Eindhoven, 1987.

[149] Malanowski, K., Finite difference approximations to constrained optimal control problems. In: *Optimization and Optimal Control*, Springer, New-York, pp.243-254, 1981.

[150] Malanowski, K., On normality of Lagrange multipliers for state constrained optimal control problems, *Optimization* **52** (2002), 75-91.

[151] Mangasarian, O.L. and Fromovitz, S., The Fritz John necessary optimality conditions in the presence of equality and inequality constraints, *Journal of Mathematical Analysis and Applications* **17** (1967), 37-47.

[152] Martinet, B., Regularisation d'inequations variationelles par approximations successives, *Revue Francaise Informat. Recherche Operationnelle* **4** (1970), 154-159.

[153] Maurer, H., Numerical solution of singular control problems using multiple shooting techniques, *Journal of Optimization Theory and Applications* **18** (1976), 235-257.

[154] Mayne, D.Q. and Polak, E., A superlinearly convergent algorithm for constrained optimization problem, *Mathematical Programming Study* **16** (1982), 45-61.

[155] Miele, A., Extremization of linear integrals by Green's theorem. In: *Optimization Techniques* (G. Leitmann, edt), Academic Press, New-York, pp.69-98, 1962.

[156] Moiseev, N.N., *Numerical Methods in the Theory of Optimal Systems* (in Russian), Nauka, Moscow, 1971.

[157] Mordukhovich, B.S., Metric approximations and necessary optimality conditions for general cases of nonsmooth extremal problems, *Soviet Math. Dokl.* **22** (1980), 526-530.

[158] Mordukhovich, B.S., *Approximation Methods in Problems of Optimization and Control*, Nauka, Moscow, 1988.

[159] Mordukhovich, B.S., Discrete approximations and refined euler-lagrange conditions for nonconvex differential inclusions, *SIAM Journal on Control and Optimization* **33** (1995), 882-915.

[160] Mordukhovich, B.S., Optimization and finite difference approximations of nonconvex differential inclusions with free time. In: *Nonsmooth Analysis and Geometric Methods in Deterministic Optimal Control* (B.S. Mordukhovich and H.J. Sussmann, eds.), Springer, New-York, pp.153-202, 1996.

[161] Morse, A.S., Pantelides, C.C. and Sastry, S., (eds.) *Special Issue on Hybrid Systems*, *Automatica* **35**, 1999.

[162] Mosco, U., Convergence of convex sets and solutions of variational inequalities, *Advances in Mathematics* **3** (1969), 510-585.

[163] Moudafi, A., Tikhonov fixed-point regularization. In: *Optimization* (V.H. Nguyen, J.J. Strodiot and P. Tossings, eds.), Springer, Berlin, pp.320-328, 1998.

[164] Nashed, M.Z., A new approach to classification and regularization of ill-posed operator equations. In: *Inverse and Ill-posed Problems* (H.W. Engl and C.W. Groetsch, eds.), Academic Press, Orlando, pp.53-75, 1987.

[165] Nijmeijer, H. and Schaft, A.J., *Nonlinear Dynamical Control Systems*, Springer, New York, 1990.

[166] Nocedal, J. and Wright, S., *Numerical Optimization*, Springer, New York, 1999.

[167] Oberle, H.J., Numerical solution of minimax optimal control problems by multiple shooting technique, *Journal of Optimization Theory and Applications* **50** (1986), 331-364.

[168] Petryshin, W., On the approximation-solvability of equations involving A-proper and pseudo-A-proper mappings, *Bull. Amer. Math. Soc.* **81** (1975), 223-312.

[169] Petryshin, W., Solvability of linear and quasilinear elliptic boundary value problems via the A-proper mapping theory, *Numer. Funct. Anal. Optim.* **2** (1980), 591-635.

[170] Piccoli, B., Hybrid systems and optimal control. In: *Proceedings of the 37th IEEE Conference on Decision and Control*, pp. 13-18, 1998.

[171] Piccoli, B., Necessary conditions for hybrid optimization. In: *Proceedings of the 38th IEEE Conference on Decision and Control*, pp. 410-415, 1999.

[172] Polyak, B.T., *Introduction to Optimization, Optimization Software*, Inc. Publ. Division, New York, 1987.

[173] Polak, E., *Optimization*, Springer, New York, 1997.

[174] Polak, E, Yang, T.H. and Mayne, D.Q., A method of centers based on barrier functions for solving optimal control problems with continuous state and control constraints, *SIAM Journal on Control and Optimization* **31** (1993), 159-179.

[175] Polak, E., On the use of consistent approximations in the solution of semi-infinite optimization and optimal control problems, *Mathematical Programming* **62** (1993), 385-414.

[176] Pontryagin, L.S., Boltyanski, V.G., Gamkrelidze, R.V. and Mischenko, E.F., *The mathematical Theory of Optimal Processes*. Wiley, New York, 1962.

[177] Press, W.H., Teukolsky, S.A., Vetterling, W.T., and Flannery, B.P., *Numerical Recipes in C*, Cambridge University Press, Cambridge, 1992.

[178] Pytlak, R., *Numerical Methods for Optimal Control Problems with State Constraints*, Springer, Berlin, 1999.

[179] Raisch, J. (1999). *Hybride Steuerungssysteme*. Shaker, Aachen.

[180] Rampazzo, F. and Vinter, R.B., Degenerate optimal control problems with state constraints, *SIAM Journal of Control and Optimization* **39** (1997), 989-1007.

[181] Reed, M. and Simon, B., *Methods of Modern Mathematical Physics. I Functional Analysis*, Academic Press, New York, 1972.

[182] Robinson, S.M., Stability theory for systems of inequalities in nonlinear programming, part II: differentiable nonlinear systems, *SIAM J. Numer. Anal.* **13** (1976), 457-513.

[183] Rockafellar, R.T., *Convex Analysis*, Princenton University Press, Princenton, 1970.

[184] Rockafellar, R.T., Monotone operators and the proximal point algorithm, *SIAM Journal on Control and Optimization* **14** (1976), 877-898.

[185] Rockafellar, R.T., Augmented Lagrange multiplier functions and applications of the proximal point algorithm in convex programming, *Mathematics of Operations Research* **1** (1976), 97-116.

[186] Rockafellar, R.T. and Wets, R.J., *Variational Analysis*, Springer, Berlin, 1998.

[187] Rosen, J., The gradient projection method for nonlinear programming, *SIAM J.* **8** (1960), 180-217.

[188] Rotin, S., *Konvergenz des Proximal-Punkt-Verfahrens für inkorrekt gestellte Optimalsteuerprobleme mit partiellen Differentialgleichungen*, Shaker, Aachen, 1999.

[189] Roubicek, T., *Relaxation in Optimization Theory and Variational Calculus*, W.de Gruyter, Berlin, 1997.

[190] Roubicek, T., Approximation theory for generalized Young measures, *Numerical Functional Analysis and Optimization* **16** (1995), 1233-1253.

[191] Rudin, W., *Functional Analysis*, McGraw-Hill Book Company, New York, 1973.

[192] Sakawa, Y., Shindo, Y. and Hashimoto, Y., Optimal control of a rotary crane, *Journal of Optimization Theory and Applications* **35** (1981), 535-557.

[193] Sawaragi, Y., Nakayama, H. and Tanino, T., *Theory of Multiobjective Optimization*, Academic Press, Orlando, 1985.

[194] Schmidt, W.H., Iterative methods for optimal control processes governed by integral equations, *International Series of Numerical Mathematics* **111** (1993), 69-82.

[195] Schmidt, W.H., *Optimalitätsbedingungen für verschiedene Aufgaben von Integralprocessen in Banachräumen und das Iterationsverfahren von Chernousko*, Thesis, University of Greifswald, Greifswald, 1988.

[196] Shaikh, M.S. and Caines, P.E., On the hybrid optimal control problem: the hybrig maximum principle and dynamic programming theorem, *IEEE Trans. Automat. Contr.*, submitted.

[197] Shioji, N. and Takahashi, W., Strong convergence of averaged approximants for asymptotically nonexpansive mappings in Banach spaces, *Journal of Approximation Theory* **97** (1999), 53-64.

[198] Solodov, M.V. and Svaiter, B.F., Forcing strong convergence of proximal point iterations in a Hilbert space, *Mathematical Programming Series A* **87** (2000), 189-202.

[199] Spellucci, P., *Numerische Verfahren der Nichtlinearen Optimierung*, Birkhäuser, Basel, 1993.

[200] Stryk, O., *Numerische Lösung Optimaler Steuerungsproblems: Diskretisierung, Parameteroptimierung und Berechnung der Adjungierten Variablen*, VDI-Verlag, Düsseldorf, 1995.

[201] Stryk, O., *User's Guide for DIRCOL. A Direct Collocation Method for the Numerical Solution of Optimal Control Problems*, Technische Universität München, München, 1999.

[202] Sussmann, H., A maximum principle for hybrid optimal control problems. In: *Proceedings of the 38th IEEE Conference on Decision and Control*, pp. 425-430, 1999.

[203] Takahashi, W., *Nonlinear Functional Analysis*, Yokohama Publishers, Yokohama, 2000.

[204] Teo, K.L., Goh, C.J. and Wong, K.H., *A Unifed Computational Approach to Optimal Control Problems*, Wiley, New York, 1991.

[205] Teo, K.L., Goh, C.J., A computational approach for a class of optimal relaxed control problems, *Journal of Optimization Theory and Applications* **60** (1989), 117-133.

[206] Thera, M. and Tichatschke R. (eds.), *Ill-Posed Variational Problems and Regularization Techniques*, Springer, Berlin, 1999.

[207] Tichomirov, V.M., *Grundprinzipien der Theorie der Extremalaufgaben*, Teubner, Leipzig, 1982.

[208] Tieu, D., Cluett, W.R. and Penlidis, A., A comparison of collocation methods for solving dynamic optimization problems, *Computational Chemical Engineering* **19** (1995), 375-381.

[209] Tikhonov, A., Methods for the regularization of optimal control problems, *Soviet. Math. Dokl.* **6** (1965), 761-763.

[210] Tikhonov, A., On the stability of the functional minimization method, *USSR Comput. Math. and Math. Phys.* **6** (1966), 26-33.

[211] Tikhonov, A.N. and Arsenin, V.J., *Solutions of Ill-Posed Problems*, Wiley, New York, 1977.

[212] Trölzsch, F., On the Lagrange-Newton-SQP method for the optimal control of semilinear parabolic equations, *SIAM Journal on Control and Optimization* **38** (1999), 294-312.

[213] Vasil'ev, F.P., *Methods for Solving the Extremal Problems* (in Russian), Nauka, Moscow, 1981.

[214] Veliov, V.M., Second order discrete approximations to linear differential inclusions, *SIAM Journal of Numerical Analysis* **29** (1992), 439-451.

[215] Walcher, S., On cooperative systems with respect to arbitrary orderings, *Journal of Mathematical Analysis and Applications* **263** (2001), 543-554.

[216] Warga, J., *Optimal Control of Differential and Functional Equation*, Academic Press, New York, 1972.

[217] Warga, J., Controllability, extremality and abnormality in nonsmooth optimal control, *Journal of Optimization Theory and Applications* **41** (1983), 239-260.

[218] Wright, S.J., Interior point method for optimal control of discrete-time systems, *Journal of Optimization Theory and Applications* **77** (1993), 161-187.

[219] Young, L.C., *Lectures on the Calculus of Variations and Optimal Control Theory*, Saunders, Philadelphia, 1969.

[220] Zarantonello, E.H., Projection on convex sets in Hilbert space and spectral theory. In: *Contributions to Nonlinear Funktional Analysis*, Academic Press, New York, pp. 237-424, 1971.

[221] Zeidler, E., *Nonlinear Functional Analysis and its Applications II/A. Linear Monotone Operators*, Springer, New York, 1990.

[222] Zeidler, E., *Nonlinear Functional Analysis and its Applications II: Nonlinear Monotone Operators*, Springer, New York, 1990.

[223] Zeidler, E., *Nonlinear Functional Analysis and its Applications III: Variational Methods and its Applications*, Springer, New York, 1990.

[224] Zeidler, E., *Nonlinear Functional Analysis and its Applications I: Fixed Point Theorems*, Springer, New York, 1990.

[225] Zolezzi, T., Well posed optimal control problems: a perturbation approach. In: *Nonsmooth Analysis and Geometric Methods in Deterministic Optimal Control* (B.S. Mordukhovich and H.J. Sussmann, eds.), Springer, New York, pp.239-256, 1996.

[226] Zowe, J. and Kurcyusz, S., Regularity and stability for the mathematical programming in Banach spaces, *Appl. Math. Optim.* **5** (1979), 49-62.